ASK
the
PLANT

Soil Fertility, Plant Analysis & Crop Nutrition

Charles Walters
& Esper K. Chandler

Acres U.S.A.
Austin, Texas

Ask the Plant

Acres U.S.A.
P.O. Box 91299
Austin, Texas 78709 U.S.A.
(512) 892-4400 • fax (512) 892-4448
info@acresusa.com • www.acresusa.com

Printed in the United States of America

Publisher's Cataloging-in-Publication

Walters, Charles, 1926-2009
Ask the plant / Charles Walters. Austin, TX, ACRES U.S.A., 2010
    xviii, 270 pp., 23 cm.
    Includes Index
    Includes Bibliography
    ISBN 978-1-60173-015-2 (trade)

    1. Organic farming. 2. Alternative agriculture. 3. Sustainable agriculture. 4. Agricultural ecology. 5. Organic fertilizers.
I. Walters, Charles, 1926-2009 II. Title.

    S605.5.W4 2010   631.58

Contents

Foreword vii

A Note from the Publisher ix

Preface xiii

1. The Plants Speak Out 1

2. Through the Soles of Your Feet — Let There Be Light 19

3. The Carbon-Water Cycle 33

4. The Carbon-Nitrogen Conundrum 39

5. The Laboratory Asks the Plant 47

6. Sourcing Fertility 81

7. What We Ask of Fertilizers 95

8. Those Variable Soils 117

9. The New Paradigm 143

10. Reading the Plant & Soil 151

11. Notes on Corn & Other Single-Fruiting Crops 169

12. Notes on Watermelons
    & Other Continuous Fruiting Crops 189

13. Notes on Citrus & Other Tree Crops 209

14. Notes on Pastures, Turf & Other Grasses 215

Appendix A: Chandler — The Making of an Agronomist 233

Appendix B: Texas Plant & Soil Lab 257

Index 259

Foreword

Earth is the only known planet to support life. A thin layer of soil covering a small portion of the earth's surface maintains all life on earth. The quality of all life on this planet is determined by the quality of that thin layer of topsoil. If we allow the quality of that thin layer to degrade, all life on earth, man included, will degrade to the same degree. The parent of all soil is mineral rock. Winds, rain, freezing and thawing break rocks into smaller sizes to start the soil-making process. Small rock particles do not become fertile soil until some life form has interacted with them.

The first life forms to attack the rock are microbes. They use elements from the air to grow and reproduce, and they slowly etch away at the rock surface. They exude, die and decompose on the rock — thus forming humus and mild acids that cause minerals to be dissolved, further enriching the accumulating soil. This process goes on and on until higher plants, and then animal life, can be sustained. The death and decay of each life has a generating effect. Each time a living thing dies and decays on the soil it creates a more fertile condition than was there before.

The energy to keep this cycle revved up comes from the sun. Plants alone have the ability to collect solar energy. This energy then passes through the food chain to all other life forms. The excrement and finally death of the many life forms allows the sun's energy to be passed to the soil to fuel the life systems in the soil and keep the cycles going so man, the highest form of life, can be sustained. Plants bridge the void between the soil and man.

Esper K. Chandler well understood the laws of recycling and return. Because of that law being ignored, he often was forced to come to the plant's rescue. Through his many years of working on farms, plantations and with other farmers, Chandler's curiosity of how plants grow, their needs and desires, filled his brain. The plants were talking to him . . . and he listened.

About 20 or more years ago an old Chevy drove up to my place one afternoon and a fellow got out and introduced himself as K. Chandler, explaining that he had heard of me, my farm and composting operation, and that he was curious. I showed him around and it wasn't long before he was telling me things about plants and their needs that weren't in any of my organic farming publications or in material from the agricultural extension service.

This gentleman's knowledge of plants fascinated me, we had a spare bedroom and I asked him to spend the night. From that point on, if Chandler was in the area, my home became his motel and our dining room became his restaurant. On each visit we stayed up late into the night talking, and it was not gossip or politics, but about plants and their needs.

Start reading this book and you too will stay up late and learn to listen to the plants.

— Malcolm Beck
San Antonio, Texas, February 2009

A Note from the Publisher

Several years ago I had the pleasure of spending some time riding around Texas' Rio Grande Valley with K. Chandler. As I saw him visit with various growers, interact with his staff, and converse with local scientists, the depth of his knowledge was obvious. But the personal trait that became screamingly obvious was his tremendous generosity in sharing knowledge. He seemed to lack the anxiety present in most of us and always had time to start at the beginning and explain how the plant receives nutrition in an understandable, practical, actionable manner to the grower. Even though he surely had taught this lesson a thousand times, his patience and interest always made it seem like the very first time.

And then there was his open-mindedness. It is easy to let emotions and personal convictions enter the fray when scurrying back and forth across the conventional-organic battle line that many activists have drawn. K. was equally comfortable consulting with organic and conventional growers alike. No matter the final prescription, he took the same approach. His goal was to provide full nutrition for the plant on a schedule that met its needs. This in turn allowed that plant to fully express its genetic potential and to best survive stress brought on by pests, pathogens or weather.

K. Chandler suffered from a weak heart (the organ, not the spirit) for many years. As he and I met with growers, virtually every one of them took the opportunity when K. was in conversation to tell me something to the effect of, "What are we going to do when K. isn't around anymore?" If you think of Texas as a country, and many still do, then Esper K. Chandler was a national treasure.

Charles Walters coupled the scientific curiosity and broad depth of classical knowledge of seekers not often seen since the Age of Discovery. But layered on top of his encyclopedic knowledge and thirst for the written word was the tenaciousness of an old-school newshound. He could crunch out copy at a pace that would make even the most seasoned of newspaper reporter stand back and take note. He loved to "get the message out."

Charles Walters retired from the day-to-day operations of Acres U.S.A. some fifteen years ago at age 67, slowing down only when his own body showed weakness in the form of macular degeneration. It is an unusually cruel trick of fate for someone who lives through and in the written word to become legally blind, but that's what happened over the next couple of years to Charles Walters. Nonetheless, he continued to write, producing hundreds of feature stories and articles for the journal he founded. Even more amazing is the fact that he researched and wrote several books over the fourteen years that he worked "in retirement."

These two old soldiers' paths crossed many times over the decades of growth of Acres U.S.A. Charles Walters knew K. Chandler well and respected his innovative methods of eco-farming. Remembering my own times with K. and the many requests over the years for us to capture his knowledge on paper, I presented the concept of a possible book to my father, Charles Walters. The goal was for them to visit on regular intervals and talk on specific topics that could be rewritten into a cohesive whole. The problem was that K. was too busy consulting to make the time to organize the vast amount of material, and Charles lacked the eyesight needed to make order out of a thousand points of insight. It appeared this very worthwhile project was to die of its own weight. It would have been a massive undertaking under the best of circumstances.

After not hearing mention of it for about a year, a box arrived with a few hundred pages of handwritten narrative. Charles Walters had written a book on the agronomy of Esper K. Chandler. The work began. The Acres U.S.A. team of Sam Bruce and Marcy Nameth handled the herculean task of getting the handwritten text into print. Anne Van Nest rescued the book in a complete restructuring and assimilation of additional materials from K. and Texas Plant and Soil Lab. Bryan

Kight did his usual magic in presenting the knowledge artistically. K. and his staff at the lab tweaked and refined the technical side. Charles Walters filled in bridges of needed narrative. Every few months K. would drift through Austin on the way north to his family farm and stop for a meeting. And then one day we declared the book complete. K. left our office with a smile on his face and appreciation for how much good this book could do, spreading these fertility and plant nutrition management techniques far beyond his own circle of farmers.

One week later we received a phone call that Esper K. Chandler suffered a debilitating stroke and was bedridden. He died a week later. We buttoned up the details of the book but were never given the opportunity to place a printed copy of this book in the hands of lead author Charles Walters, as he passed away a few months later as well. Both men lived long, full, productive, rich lives. Both men sought to leave the world in a better place than they found it. It is with particular pride that we present the life work and testament of Esper K. Chandler as told by Charles Walters. Both men would want nothing more than for a new generation of farmers and consultants to build upon this knowledge, eclipse it, and move agriculture to an even more sophisticated place. They would both dream of a place where nutrition rules and toxic chemistry is laughed at and known to be unnecessary.

— Fred C. Walters, Publisher
Austin, Texas, November, 2009

Preface

A few issues after *Acres U.S.A.* was launched, I encountered the work of Clive Backster. It landed on my desk as an entry in *Argosy* magazine and from other sources. The arresting introduction said that if plants could scream like pigs we couldn't stand the noise in the countryside. The report went on to recite the conventional wisdom about some 92 elements, pointing out that few were required, others were needed in very small amounts. It was hunger that caused that silent pain, and toxic rescue chemistry exacerbated that pain, a proposition that researcher Clive Backster sought to prove with his lie detector tests. Backster relied on the advice of botanist Barbara Pickard that plants give off electric signals much like nerve impulses in animals. These signals produce a pattern that resembles an electrocardiograph recording. This electricity may set off hormones that are released when a change occurs in the environment or when something disturbs the plants.

Pickard was a Washington University scientist when her work counseled Backster. Their joint conclusion was that plants could scream like pigs, albeit in silence. More important, if we could only hear them, they might tell us not to starve them or shower them with harsh pesticides of every stripe.

Plants are directed by nature to develop hormone and enzyme systems to protect themselves from insect, bacterial, viral and fungal crop destroyers. Clive Backster encountered some of these facts when he introduced the lie detector into the fray. After assembling a database, he was forced to conclude that even though plants could not squeal out loud, they had means of telling us of their food deficiencies, which we speak of as hunger signs. Of the 92 non-radioactive elements of which the earth is composed, plants seem to require considerable amounts of perhaps ten, the so-called major elements, this according to early research. Of the rest, only minute amounts were believed to be required. Later research revealed a role for each of the residents on the Mendeleev periodic table of the elements. In the pages that follow, a lot of attention will be given to hydrogen, oxygen and nitrogen — all of which come directly or indirectly from water and air — and phosphorus, potassium, calcium, sulfur, magnesium and boron, which are sourced directly from the soil.

It was Backster who first discovered that plants could trace out emotional reactions when electrodes were applied while watering. The polygraph even recorded Backster's thought, "I'll burn the plant." A paper was published on these findings in the *International Journal of Parapsychology* in 1969.

In one of the experiments, a man was selected by lot to kill a plant: pull it from the pot, mash it, murder it in the first degree. Backster had no difficulty finding the "murderer." He simply attached the electrodes to a "witness" philodendron. When the suspects were marched into the room one at a time, Backster got a reading when the killer entered.

Incredible? Perhaps. Perhaps, too, there really is such a thing as a green thumb. Backster claimed that plants either like a gardener or they don't. How well do plants like an insensitive agronomist, one who showers the environment with harsh salts and even harsher chemicals?

Why do I tell this hard-to-believe tale? It doesn't take such extraordinary measures to detect a plant's hidden hungers — or hungers to come. The laboratory can tell us. And the fantastic of yesterday has become the reality of today in the person of Esper K. Chandler, agronomist, laboratory proprietor, seeker after nature's secrets. Chandler hasn't entered the fertile fields of the Rio Grande Valley with a lie detector in tow, but rather with a unique understanding, interpretations and synthesis of soil, leaf and petiole tests. This agronomist has taken seriously the proposition that plants can communicate. His birth certificate name is Esper Kaylor Chandler, but the world of sustainable farming calls him K. He's as lean as a creekbank willow and, as one of his clients put it, "he has smarts."

Malcolm Beck, the founder of San Antonio-based Garden-Ville has called Chandler a national treasure. Malcolm tells of a conversation he had with Joe Bradford, an agronomist at the USDA-ARS Kika de la

Garza Research Center in Weslaco, Texas. They agreed that some way had to be found to download the mind of Chandler so that present and future generations could avail themselves of his insight and his philosophy: "ask the plant." To seek answers is to imply knowledge necessary to frame the right question.

Chandler is the proprietor of a leading soil testing lab that understands and makes good natural/organic and sustainable, as well as conventional, test interpretations and recommendations. "You're always talking about asking the plant," one client commented, confronting Chandler. "Mr. Chandler, what language do you use to talk to and listen to your plants?"

Chandler's answer was slow and deliberate as befits a son of the South. "It's sign language," came the soothing answer. He went on to explain that "we have to learn a new language when we are growing plants. There are many types of sign languages."

It was an opener that Chandler has been required to recite countless times his career was propelled into a lead role in the movement styled "eco-agriculture" as far back as 1971. In fact, all of the nomenclature used to define biologically correct farming means essentially the same thing. One and all, the natural/organic farming systems expect plants with access to balanced nutrition to build their hormone and enzyme systems so they can mount their own defenses against crop destroyers and weed proliferation. The order is tall, and not all attempts to comply with nature's demands are successful.

"The main sign language that we look at can be compared to the standard blood test that physicians use to diagnose the human situation," Chandler tells his clients.

With these few whiplash lines Chandler takes the client, the student, the fellow professional into the anatomy of plant life. The plant analysis in itself invites criticism simply because it has only a spot validity for a life cycle that changes the metes and bounds of signs and systems at an almost reckless pace.

Once that plant pokes its head above the ground, the scene changes, often by the hour, sometimes in less time than that. "Deficiency symptoms tell the grower about nutrient problems somehow left out of the rootbed calculations." With that statement Chandler opens up a pantheon of minerals that ought to be but aren't present.

In the pages that follow, Chandler will explain what flowers and weeds are, and why the petiole preaches to the farmer with warnings once reserved for a march down the church aisle to the mourner's bench.

You will find a measure of quoted material in the narrative pages of this work. These quotes are real, not reconstructed or fictionalized. The tape recorder came to my assistance while I was trying to download the mind of Chandler. Other documents used to construct this manual of a living agronomist are cited as well.

Withal, Chandler is no academic purist who farms with a pencil a thousand miles from a sugarcane field or cotton patch. He walks the acres under his tillage with a keen eye. If a sheep-nose citrus fruit hangs there in the open, Chandler will recite what went wrong and what can be done today, at this hour of the growing season.

Ask the plant! It will tell the grower whether the soil is too wet or too salty or too dry. You can work your way through a hundred texts on plant and crop production without encountering leaf analysis, petiole reading — yes, *reading* — or a tissue test. Even when made a matter of reference, "Too often," reminds Chandler, "any type of test on a plant is referred to as a tissue test." Chandler insists on being more precise. "The tissue test," he reminds, "has to do with chlorphyll, sugar content. Or quick sap tests in the field on freshly removed portions of the plant. These are tissue tests."

It might well be generalized that most farmers are still seat-of-the-pants farmers. They do not test soils. They certainly don't walk the fields with a fishing-tackle box replete with paraphernalia needed for on-scene testing. The heads-up farmer nowadays has to finish growing the present crop in an economical and profitable way.

And a short inventory of information on the petiole approach reveals insight you hardly hear about either in the classroom or in the extant literature. Petiole testing involves the sap of a plant's leaf stem. This is done by selecting the most recent fully expanded leaf on the plant. The stem, or petiole, has to be separated from the leaf connected to the main stalk or vine of the plant.

Leaf analysis has requirements all its own. The most recent fully expanded leaf has to be selected. That leaf is no longer taking any-thing out of the "bloodstream" sap — if that term can be used — such as transferring nutrients from older leaves to younger leaves. "If you detach the older leaf, you won't know if all the nutrients present are from the supply deriving from the root soil, or from foliar applications, or from the transfer of nutrients."

The few notes presented here merely declare that a new way of thinking is here to stay. It will require cancellation of some few dearly held beliefs and not a little "settled" science. Indeed, many settled sciences are challenged as Chandler unveils many of nature's secrets. Not least is the idea of giving the soil a nutrient fix good for the season. Like a youngster on the grow, plants have their recommended daily allowance, and it had better be there on time if top performance is to be expected.

It may be an over-simplification, but it is probably correct to say that in addition to soil testing four types of field tests telegraph everything the laboratory has to say: the tissue test, the sap test, the petiole test, and the content of the leaf analysis. The ramifications are precise and awesome, for which reason each will require exposition and analysis as these chapters unfold.

The Rio Grande crops and those of the nearby Deep South are numerous, a fact that removes Chandler's laboratory and consulting practice from the mainstream. It is no accident that Chandler can answer almost every question about alfalfa, peanuts, cotton, sugarcane, corn, every Rio Grande Valley vegetable and citrus crop, and crops indigenous to the High Plains, the northern tier of states, with the Ohio Valley and New England thrown in for good measure.

Each crop has its own character. If it bows down or loses its posture and color, this is not subservience, but an appeal for help. Ever since Rachel Carson, the struggle has been difficult for those trying to turn agriculture around. "But people like Chandler are winning," according to Malcolm Beck. The world may prattle about global warming, but leaders of the movement like Chandler know that the remedy for environmental ills is seated squarely in the thinking apparatus of those who keenly observe nature.

It is strange that grants seem to be available for research into whether dogs bark with an accent or hawks make their lazy circles in the sky turning left or right, yet there is a paucity of foundation interest on more important nuances of living plants that provide *Homo sapiens* their food. Stranger still, most of the recent discoveries have been made by innovators in the field, not in academic circles.

Here then is the download of Chandler's mind and activity, a paean to a modern pioneer.

— Charles Walters
Raytown, Missouri, September 2008

The Plants Speak Out

It takes a cloudless night an adequate distance from the city's light pollution to really appreciate the beautiful planet on which we live. Telescopes can take us well beyond the Milky Way, yet the unaided eye can find some planets in our solar system, and a little book learning can supply the intelligence that we have been here some 14 billion years, more or less.

This organism called Earth is no more than a speck in our planetary system, one that is swung on a gravitational string in a 300-million-mile orbit around a nebular sun. It wobbles slightly on its axis so that each hemisphere can be blessed with summer, winter, spring and fall.

Geologists tell us that planet Earth has many more mineral compounds than our sister planets, all of them fashioned from those elements that are blocked with such orderly symmetry on the Mendeleev chart.

How can these minerals have evolved from the same elements that service other planets? This evolution of inert minerals is aided and abetted by the life forms called microorganisms. Our microbial workers did not raise mountains from the deep. A fiery heartbeat from the center of the earth did that, striking land masses with tsunamis, sending water up or into a frigid air envelope, igniting ocean warming and great ice ages.

Some four million years ago, humanity appeared on the living planet. There developed, we reckon, a symbiotic relationship between life forms, with man asking the plant, and the plant handing off answers.

Those minerals that floated in on the air or migrated through the soil asked for their blessings from the microbes, either in the sap of the plant or by the soil's billions of unpaid workers.

Even Texas seems puny under an umbrella of stars. Quite recently astronomers informed us that discernable planets orbited a star well removed from our solar system, and still credentialed and settled science suggests that the elements of the universe are fixed. Some few elements fall from the heavens, but for the most part planet Earth's supply is fixed. That supply services not only bacterial fixers, but also plants, animals and human beings.

The high plains in the Texas Panhandle, the hills west of Austin and San Antonio, the undulating grasslands, the black soil corridor, the rainbelt to the east that merges with Louisiana's sugarcane acres, and cropland around the world all challenge the food producer. None of them, however, field the diverse challenges of the Rio Grande Valley.

Here, Esper Kaylor (K.) Chandler, proprietor of the Texas Plant & Soil Laboratory, modestly wears a reputation that brings agronomists from across the nation and around the world to his digs. Chandler asks the plant and teaches farmers its language and how its answers give the crop and its end-line consumers permission for life. He walks a razor-thin line, counseling quality when an industry asks only for bins and bushels.

Chandler is the daring adventurer whose downloaded knowledge will, I believe, annihilate the existing order. Sir Albert Howard counseled that two false premises — partial and imbalanced fertilization and toxic rescue chemistry — have invaded the republics of learning. Chandler's critiques have been tested, and they sting like a driven nail as the plant tells agriculture to learn its language. The lessons contained therein start right now.

Come walk with me, or ride in my pickup truck; learn how to see what you look at. Meet the growers and their consultants and the plants that tell them what to do. James "Bubba" King and visiting plainsmen will tell you they resonate with what Esper K. Chandler knows and, more important, what the plant knows.

"Factory calculated fertilizers are made soluble," says Chandler, "and it's a mixed blessing. As Albrecht says [William A. Albrecht, former head of the Department of Soils, University of Missouri], they should be insoluble but available." In a drip line or sprayed as a foliar, the discourse goes on, "It's up to microorganisms to make them ready for plant uptake. What we worry about is the plant going hungry before we know it."

So-called conventional agriculture pretends to know all the answers. Its absolutes reach back to the days of Justus von Liebig, the "Father of modern fertilization," who found the connection between nitrogen and plant growth and defined as his absolute the primacy of that element. Albrecht Daniel Thaer said absolutely that humus was crown prince if not king, and Robert Lawe proved absolutely that nitrogen, phosphorus and potash, N, P and K, were the essentials best made factory soluble for field use. Lawmakers have since enshrined NPK as legal writ, complete and settled.

"And nature smiled a sphinx-like smile," wrote a poet, "and let them have their way, and waited patiently for the fools to go away!"

The pickup rolls past a bat cave.

A Lesson About Fertility from the Lowly Bat

Chandler had his experience with phosphoric acid as manufactured and under the auspices of bat guano. "When I was mining bat guano in the caves, we were after phosphate naturally made," Chandler recalls. "Those bat caves have limestone cathedral ceilings, which is where the bats spend daylight hours. Trickles of water penetrated the limestone from the soil above. Ammonia rises from the fresh guano and hits the lime water. Ammonia water then cuts into the limestone, and fallout peppers the fresh guano on the floor of the cave. Some falling material is the size of an automobile. As the limestone is showered with the flush of fresh droppings, composting goes to work. A veritable mountain of material develops over time. Texas bats consume billions of insects even in the upper atmosphere, intercepting predatory insects heading to northern and Midwestern fields.

"We took off the fresh manure pile to reach an early formation of rock phosphate. As those phosphoric acids reacted with the calcium, large rocks would fracture with the tap of a sledgehammer. A bit deeper, the product to be mined is truly rock phosphate," Chandler explains. "That's what we mined and sold. And we used the petiole test to evaluate the uptake of guano phosphates by plants." This was the start of Chandler's career in plant analysis consulting services under the direction of Dr. Albin D. Lengyel, Phoenix, Arizona, at the request of local Winter Garden Farmers in Texas.

Organic standards, to be explained later, do approve of hydrogen peroxide. They don't quite approve of urea. Is phosphate produced by bat guano acceptable? Does nature approve? Yes!

Growers who accept a whole battery of organic products hold open the final answer, pending the arrival of new databases that will be rejected or ratified as the organic movement asserts itself, and as the new innovators assert themselves, always asking, "Does nature approve?"

Nature's way decrees that plants love company, not necessarily their own. A pasture that appears to be a monoculture will have, say, 40 to more than 50 species. The tomato plant treasures sweet basil, probably because the two plants trade auxins to their mutual benefit. Academia has largely ignored this strange plant alchemy, but gardeners have asked questions for centuries. They might better have asked the plants. Chandler takes note of what talented amateurs have discovered, for which reason he is interested in area gardeners who have bush beans living peacefully with chard and beets, and corn trading auxins with almost every member of the bean family.

Weeds are an index of what is wrong with a soil system, and sometimes what is right. Our earlier book *Weeds — Control Without Poisons* attempts to plumb the range of this practice, secure in the knowledge that Chandler has been there before. Certainly, the plant that arrives first declares the kingdom for the year.

Reading the Plant

It is not practical to have companion plants in most field situations, but it is possible to read deficiencies and surpluses of plant nutrients even without the certainty of a petiole test, a refractometer readout, or a laboratory's final audit.

Plant Analysis Guide Sheet

Nitrate (NO$_3$ in ppm) — Nitrates in the sap represent future growth. Their effect is visible in 10-14 days. Too many nitrates too soon reduces fruit set. Changes in nitrates take effect fast.

Phosphate (PO$_4$ in ppm) — Phosphates are in the sap for future growth. They show present root activity. Phosphates can be increased with humus, plant growth regulators and microbes.

Potassium (K %) — Potassium affects water uptake and efficiency, sugar production, enzyme formation and plant health. There is a high potassium requirement to form sugars.

Sodium (Na %) — Low amounts of sodium are best; just a trace is essential.

Calcium (Ca %) — Calcium is needed for cell walls, nitrate utilization, roots, leaves, pollination, and development fruit set.

Magnesium (Mg %) — Magnesium is needed for chlorophyll, photosynthesis, phosphorus metabolism and respiration.

Zinc (Zn in ppm) — Zinc is a plant growth stimulator. It affects enzymes, and metabolic reaction.

Iron (Fe in ppm) — Iron is needed for respiration, chlorophyll formation and energy. It is an oxygen carrier.

Manganese (Mn in ppm) — Manganese is needed for enzyme activation, photosynthesis, and maturity. It is connected with P and Ca.

Copper (Cu in ppm) — Copper is needed for chlorophyll formation and energy. It also catalyzes plant functions.

Boron (B in ppm) — Boron is needed for nitrate uptake, calcium utilization, pollination and sugar transport.

Notes About Future Plant Development

a. Foliar applications of nutrients have little or no immediate effect on the sap as they stay in the leaves to aid plant functions. Micronutrients do not translocate like NPK, which can transfer from old to new leaves when sap supply is deficient, not so with the micronutrients Zn, Mn, Fe, Cu, B, Mo, etc. Ca and Mg seldom move very little if at all.

b. Low micronutrients in the sap show the need for foliar applications and/or soil amendment.

c. Foliar micronutrients applied on the leaf do not show in the sap.

d. New leaves will continue to need micronutrients until sap supply improves.

e. Weekly foliar feeding will be needed every 5-10 days. Remember plants feed every day!

Nitrogen (N) is a problem when too much is used too early. It reduces uptake of other nutrients and contributes to disease and insect attack.

Even heavy fruiting plants can only utilize about 10 lb./acre of actual N per week.

Less than 20% of this N is needed during the first 6-8 weeks of growth (more for grain).

Ask the plant and feed when and only what is needed. Apply it in small increments added to the soil or as a foliar spray where possible.

Phosphate (PO_4) in the sap shows root activity. Phosphorus is mostly taken up by young root hairs near the growing tip of roots.

Older roots show slower uptake, which means senescence or cutout is occurring.

Roots can be stimulated with humus products, multi-hormones, biologicals, enzymes, etc.

P availability is helped by the availability of S, Ca and other natural materials.

Sudden changes in P uptake can be the result of new root growth interruptions caused by too much or too little water and/or lack of P, cultivation, blight, compaction, nematodes, disease, etc.

Maximum economic yield requires weekly testing for a crop-logging program.

Before you get out of the pickup you can see that the plant has a calcium deficiency. The leaves are standing very erect. One of the symptoms of calcium deficiency is corrugated leaves, like sheet iron. Adjusting for material, they bend about as easily as sheet iron. If the leaves bend at all, they break. This, when a leaf should have a good sweeping curve in it. Calcium, having low mobility often appears first on older leaves. Classic symptoms appear as blossom-end rot of tomatoes, tip burn of lettuce and dead tissue in rapidly growing leaf areas.

In slower growing plants, calcium is moved from the older leaves to the younger ones. This results in the margins of the leaves growing slower than the rest of the leaf and the leaf cups downward. In severe cases the petioles develop but the leaves do not. Plants growing under chronic calcium deficiency will wilt before non-stressed plants.

DEFICIENCY	SYMPTOMS	COMMENTS
Calcium (Ca)	Corrugated, distorted younger leaves, early wilting, upward, erect orientation, leaves break before bending. Necrotic areas (dead tissue) in rapidly growing leaves. Leaf margins cup downward. Severe cases — petioles form without leaves. Shows as blossom-end rot on tomatoes, and tip burn on lettuce.	Non-mobile. Deficiency first appears on younger leaves and then progresses to older leaves.

Look again, this plant has a phosphorus deficiency. The green is too dark. The photosynthesis is moving out of the leaf. Phosphorus is the transport element that performs this function. The plant has spoken: excess nitrogen here, lack of phosphorus there. The leaf is an indictment that simplistic nitrogen has been applied, an inherent imbalance has developed. Visual inspection next kicks open the door. Excessive nitrogen is nothing but a waste. Nitrogen, being a water-soluble element, leaches away and is bad for soils and the environment. Adequate nitrogen is essential for productive plant growth and fruiting. Too little and plants are yellow and stunted with smaller flowers and fruit. Too much and lush plants produce few flowers and fruit.

DEFICIENCY	SYMPTOMS	COMMENTS
Nitrogen (N)	Yellow color starting with older leaves. Young leaves are a light green color and become progressively smaller. Stunted plants with smaller flowers and fruit. Extreme deficiency results in early wilting.	Deficiency appears in older leaves first. Excessive soil water and leaching will remove nitrogen from the soil. Symptoms reverse quickly when treated.

SURPLUS	SYMPTOMS	COMMENTS
Nitrogen (N)	Lush growth with excessive foliage. Few flowers and fruit. Overly green color. Thinner, weaker stems. Increased pest and disease incidence.	Very mobile in the plant and soil. Excess nitrogen is detrimental to the soil and environment. Water soluble and prone to leaching.

Phosphorus deficiency symptoms are disparate, not distinctive and difficult to identify. Often plants are dwarfed or stunted and can be mistaken for much younger plants. Many plants, including tomato, lettuce, corn and brassicas have a purple color to the stem, petiole and undersides of the leaves. If severe, young leaves are blue-gray, while older ones have a brown veining. Cold weather can cause a temporary phosphorus deficiency.

Chandler comments, "You get striping on corn and small grains, as interveinal chlorosis. That purple color in corn denotes a phosphorus deficiency. The purple color is widespread, not just in one spot. The sign can trim the edge of the leaf or take it all."

DEFICIENCY	SYMPTOMS	COMMENTS
Phosphorus (P)	Dwarf, stunted plants. Delayed growth and smaller plants. A purple or darker green coloration to older leaves and stems. A red tinge to veins. Interveinal chlorosis and leaf tip burn on some crops.	Difficult to identify. Cold weather can cause a temporary phosphorus deficiency. Can be locked up in very acid or alkaline pH levels.

Chlorophyll production and respiration are important plant functions that require copper. Usually, crop yields fall significantly before any visible deficiency signs appear. A copper deficiency is revealed by the pointed end of the leaf on onions, sugarcane, corn or grain sorghum. Leaves are curled, and their petioles bend downward. The deficiency causes a unique corkscrew or pigtail. It dries out as if to construct a corkscrew for a wine bottle. The dryout in most plants turns bleached gray.

DEFICIENCY	SYMPTOMS	COMMENTS
Copper (Cu)	Reduced crop yields. Pointed leaf shape on some crops. Curled leaves and downward petiole. Bleached leaves and early wilting. Stunted plants with dark green leaves and stem dieback.	More prevalent in sandy soils with low organic matter. Can be less available in soils at higher pH levels.

Iron deficiency exhibits interveinal chlorosis on the whole length of the leaf. Since iron has low mobility, the symptoms start with the youngest leaves and progresses until the entire leaf is totally bleached. An application of iron will reverse all but the worst deficiency conditions. Iron deficiency is more common with anaerobic calcareous soils and in association with excess heavy metals.

DEFICIENCY	SYMPTOMS	COMMENTS
Iron (Fe)	Light green, yellow to white interveinal chlorosis that progresses until the entire leaf is bleached. Youngest leaves show the first symptoms.	Low mobility in the plant. Less availability in soils at higher pH levels.

Zinc deficiency also reveals interveinal chlorosis, but in the basal part of the leaf. Zinc is immobile in the plant and in early stages affects younger leaves that become yellow and pitted between the veins. In many plants, leaves become smaller and internodes shorten. Guttation is often visible.

DEFICIENCY	SYMPTOMS	COMMENTS
Zinc (Zn)	Interveinal chlorosis, first at the base of the leaf. Younger leaves show first symptoms of yellowing and pitting between the veins. Smaller terminal leaves may become rosetted with shorter internodes.	Low mobility. Deficiencies are more prevalent during times of low microbial activity in the soil. Less availability in soils at higher pH levels.

Manganese deficiency is characterized by a different interveinal chlorosis, this time in the upper half of the leaf. Early symptoms are similar to iron deficiency. They begin with a light chlorosis of the young leaves and netted veins on older leaves. As deficiencies continue, plants have a gray metallic sheen and purplish luster on the upper leaf surface. Grains are particularly susceptible to manganese deficiencies. Here, eventually the entire leaf withers and dies.

DEFICIENCY	SYMPTOMS	COMMENTS
Manganese (Mn)	Interveinal chlorosis, starting with the tip of the young leaf. Similar to iron deficiency. Gray purple coloration. Delayed maturation and white necrotic leaf spots can be seen on some crops.	Less availability in soils at higher pH levels.

Potassium deficiency shows first in the older leaves. They wilt; look dry, leathery and scorched on the margins, and if not reversed, symptoms progress inward between the veins and yellowing becomes necrotic. Leaves of some crops can curl and crinkle, leaf tips may die. Severe deficiencies result in plant stunting and reduced yield. A higher incidence of disease and a stunted root system also are symptoms of potassium deficiency.

DEFICIENCY	SYMPTOMS	COMMENTS
Potassium (K)	Yellow margins around older leaves that turn tan-white and eventually look scorched. Severe deficiency shows on younger leaves. Some crops develop white flecking on the leaf blades or purpling under the leaf. Uneven fruit development.	Mobile in the plant. Less availability under lower pH (acidic) soils.

Calcareous soils are as unforgiving as the micronutrients they back up or do without: iron, zinc, manganese and copper. All subject plants endure the shortfall.

The pickup truck can take you from field to field, but it is not a spectator's conduit. You have to feel the give of the soil under your feet to know the soil, and you have to tryst with the leaves to get their messages.

Carrots are hard to read because the leaves are so small. Usually it takes a severe problem before a silent scream becomes evident. Then you can see a phosphorus deficiency because the tops are too dark. Pull a sample and find a split root, either a bifurcated split or a lateral split: boron deficiency. If the crack takes place early in the growth it will widen out, often to three quarters of an inch wide. The carrot has two sections: the outer section and the core. Usually the crack is limited to the outer section.

All states have potato acres, and all potatoes tell of their nutritional status if asked. The most common telltale sign speaks of a manganese deficiency. The sign is common to dicots, namely interveinal chlorosis, although the area near the vein still stays green. In the case of zinc or iron deficiency, the complete inner vein area will be yellow or white as chalk, only the vein will remain green.

There are corn and grain sorghum acres in the Rio Grande valley of Texas. The symptoms speak to the inter-relatedness of plants for they are somewhat the same as those exhibited above. Soybeans present their leaves for inspection, exhibiting chlorosis in a part of the inter-veinal area. Here the dicot rule bends, and areas just off the chlorotic veins are painted green. Further out the green turns white.

Hold it a minute. Let's take a look at dicots as a class. "I look at what plant part is affected," Bubba King, a leading expert in listening to the plants' sign language tells Chandler as the two compare notes that are as esoteric as the key to an enzyme. Is all or part of the plant affected? Is the plant new, old? Is it the mid-rib or the innerveinal area? Are the colors deep green, pale, red, purple, orange-yellow, brown, tan, white, or black? Purple means calcium and phosphorus deficiency. Colors can identify the status of a plant when they denote starvation.

The two men move on. King has been reading plants ever since he first arrived on the scene working on behalf of growers stressed by both markets and the demand to produce more efficiently. Chandler made the connection between plant, soil, and laboratory early on as well.

There are leaf patterns that suggest discordant music. There is a Christmas tree pattern. It looks like a fir tree. "You're losing chlorophyll," intones King. The mid-vein supplies the trunk, and service veins make up the branches. The outer area stays green, the rest turns yellow.

Leaves are only an index. There is the overall plant. Is it stunted, spindly, spotted? Are the leaves undersized? Narrow? Misshaped? Cupped or wrinkled or rolled? Are the leaves brittle, cracked, split, corkscrewed, or do they exhibit lesions? All reveal a boron deficiency. Even if a plant stores plenty of boron, it does not translocate in the plant; the nutrient isn't making it to the petiole or the leaf. Boron figures in cellular elongation, the metabolism of calcium, tubular growth, nitrogen transport, sugar movement, photosynthetic transport, and subtle effects on the fruit.

What is the choice for windbreaks in the watermelon fields? Mostly, small grains are used. "You don't have to leave the pickup," King tells Chandler. "The scream is that loud." These plants will stand erect, then break off because of calcium deficiency. "Magnesium is also a major deficiency," says Chandler.

Magnesium deficiency shows up in the older leaves of the plant because that mineral is somewhat mobile. It can be pulled out of older leaves and put into new growth. "Slowly mobile," adds Chandler. The rolled leaves lose their chlorophyll, turn yellow or sometimes achieve an orange tint. At times the orange will turn pink to reddish. This abridged rainbow of colors is very distinctive of magnesium deficiency.

DEFICIENCY	SYMPTOMS	COMMENTS
Magnesium (Mg)	Yellowing between leaf veins starting with older leaves. Leaves loose chlorophyll and turn from yellow to orange, pink or reddish. Veins may stay green. Leaves roll and fruit may be small and woody.	Somewhat mobile in the plant. Can be confused with other nutrient deficiencies (especially iron) or virus symptoms.

Field signs and observations should accompany samples for petiole analysis, but often they don't. When they do, the laboratory is several strides along the path of offering consultation that can rescue a crop or at the bare minimum devise the economic balance of production to the packing shed.

When triticale is used as a windbreak, the crop has the earmarks of wheat, good or bad. Ditto barley. Notes are helpful on the information sheet that accompanies a petiole into the laboratory, but in fact a sampling of leaves will serve as an adequate designation. Often the same elements prevail. Citrus trees, pecans — dicots all! One of the symptoms of calcium deficiency is cupped leaves, up or down. Often the leaves become boat-shaped. This baffles people who expect surplus calcium in calcareous soils. Now that unavailable phosphorus rears its head. Many labs fail to designate available and unavailable, which is a Texas Plant & Soil Lab specialty. There is a sameness about signs and symptoms, whether oranges, grapefruit, avocadoes, pecans — actually all broadleaf plants.

Evergreen trees are hard to read because needles do not follow the rules of leaves. Loss of needles often arrives, the necessary consequence of nutrient starvation.

A to Z, the inventory of plants that come under the scrutiny of Chandler's keen eye include artichokes, beets, all vegetable crops actually, cotton, corn, sugarcane, grain sorghum, all the small grains, and citrus. It is safe to say that only the movement of nutrients to the leaf and fruit really counts.

Root crops behave like carrots when boron is missing or inadequate. Beets probably need the greatest amount of boron of any commercial crop. Just the same, the table beet will grow in high-salt soil.

All totaled, the Texas Plant & Soil Laboratory presides over 65+ crops, including cranberries and pomegranates, whatever the grocery store offers. Chandler can even grow bananas, not a Texas crop. There is a reason for this. Any consultant can learn valid lessons from banana plantations, their cultural practices, record keeping, and history. Unfortunately, petiole analysis and natural methods are not often a part of banana production practices. They pour on harsh fertilizers, but their yield goes down, down, down.

A week's travel through the Rio Grande Valley and into Mexico and Central America serves up knowledge not dreamed of by people who till with a pencil and stay miles away from the field. The official book on coffee — not a mainland U.S. crop — says phosphorus does not figure in coffee production. The problem is that it does. Under normal conditions in coffee production, soils are quite acidic. Phosphorus added to aid uptake contests with aluminum and iron. If the pH is adjusted, coffee takes up the phosphorus it needs.

The plant proposes and it is up to the grower to dispose. This is basic to the agronomy and wisdom of Esper K. Chandler.

The Education of an Agronomist

Some 40+ years ago Chandler and his partner Albin Lengyel both talked the talk and walked the walk with farmers, bringing advanced knowledge to the field where it counted. Lengyel came to the U.S. to escape Hitler. He joined the Navy, and after the war went to school at Maryland and Purdue, taking degrees in agronomy, plant physiology, and biochemistry. The other outstanding colleague who rated the recognition Lengyel deserved was an Arkansas professor named Stutte.

Ph.D.s Albin D. Lengyel and Charles A. Stutte "counseled me into plant analysis and physiology," Chandler recalls. It took giant strides to keep up with those two practical schoolmen. In those days, solid agricultural knowledge was far removed from a lad who grew up on a Louisiana farm with worn-out land. A decade-long depression told those who would listen that they might as well banish thoughts of higher education. The crop was bumblebee cotton. "It wouldn't even feed a bumble bee, much less a boll," Chandler remembers.

"We lived on what we grew. Louisiana State University said there was no future in cotton in that area." Just the same, Chandler's sharecropper, tenant farmer, and finally owner of Grandfathers' 180-acre hardscrabble farm allowed subsistence.

It was a fertilizer-free and chemical-free agriculture enterprise of necessity. The family couldn't afford the former and the chemical industry hadn't yet developed the latter. Dairy animals were maintained almost as much for their manure as for their milk. "We composted and didn't even know it by picking up cow pies from the milk cow lot each morning and placing them in a bin," Chandler remembers. "We asked the plant to some extent, and we didn't know that, either." Those eight to 16 milk cows fertilized the farm. There was little market for cotton seed, so it was returned from the gin and fed to the cows. Any leftover fertilized the crop. Crop rotations and legumes were also used.

Fast forward 20 years. A LSU research plot by Chandler and others on a farm next door made $2\frac{1}{2}$ bales of cotton on each of its 40 acres compared to the family's Depression crop of half a bale per acre. That got Chandler's attention, and he has been asking the plants ever since. If records still existed, they would tell of 40 acres being given the intensified fertility treatment replete with chemical controls, nematocides, and insecticides, all to make two-and-a-half bales. The initial response was as follows, this while talking to the Extension agent and Chandler. The farmer Archie Davis said, "If you'd have told us this 20 years ago, we'd still be farming in this parish instead of producing pine trees."

In its microcosmic setting, chemical agriculture ran its course quite early. Pine trees became the best answer, and early forestry fertility research that Chandler participated in nailed down phosphates as the fertilizer from which response could be expected. N and K were not important factors in producing pine pulpwood and saw logs. During those transitional years, farmers still producing cotton used 4-8-4,

5-10-5, or 6-12-6 as high-analysis grade controls were enforced on fertilizer. With phosphate uptake no more than 10 to 15 percent, the soil colloids were loaded with single superphosphates and gypsum when the pine trees arrived. The basic low-analysis fertilizers were mainly composed of 18 to 20 percent single superphosphate, which contains 55 percent gypsum, 18 percent calcium, and 12 percent sulfur before Eisenhower's Soil Bank Program demolished small farms in the South.

History, experience, even growing up, all pointed to plant analysis, and plant analysis came to mean "questioning plants." The bottom line, according to Chandler, is that the farmer wants the best uptake of plant nutrients, whatever the source. In order to give real efficiency to that uptake problem, "we have to reestablish the humus function of the soil, the basis for natural/organic sustainable farming."

The drip-irrigation system is a valued key. Not only does it make it possible to feed the plant vital nutrients, especially phosphates, on a cafeteria basis, it enables uptake of existing soil phosphates, thereby doubling the potential. Then multiple products of humus, lignosulfates, enzymes, soil inoculants, hormones such as in seaweeds, cytokinins, auxins, and gibberellins bring on another synergistic effect, all of them answering the plant's call for help. Agronomists have had to reassess the role of nature's unpaid workers, soil microorganisms, and their role in forcing a new look at the observed facts of the situation. Over the past 40 years, organiculture has kicked open the door to new thought processes and reopened old ones. Chandler is the agronomist who most clearly achieved consensus between natural/organic, sustainable, and conventional farming. He will tell you about Dan Logan when the occasion is right.

"He was the first guy to hire me out of college. He started me off as a tractor driver and nutured my moral, educational and management capabilities. He was practical. He was a leader in the cotton industry. He understood more plant physiology than a Ph.D. We'd walk the fields every week. My job was to monitor and culture the crop. He said, 'K., if you just walk across the field, the knowledge is just going to seep up through the soles of your feet and through the seat of your pants. That plant is just going to tell you. You're going to know it. I can't write the language for you, but by walking, looking, and listening to the plants, you'll learn what's going on.'"

Why is this book called "Ask the Plant?" Because the plant is the judge, jury and executioner. A healthy soil needs to be present for a healthy plant. Asking the plant is the best way to find out what nutrients are being taken up by the roots at a specific time.

Through the Soles of Your Feet
Let There Be Light

Even before the Earth came to position itself at a distance that enabled life, the injunction of Genesis 1:3, "Let there be light," was answered. There was light. It came by an undulatory, vibrational movement at a velocity of 186,282 miles per second. Sunlight, with its subatomic particles, its finely rationed, lethal yet necessary quarks, has been the first principle of existence ever since orbits and spins defined seconds, hours, and days.

Light also happens to be the first principle of health for plants, animals, and humans. Without it not a blade of grass, tree, wildlife, or domestic animal could grow. It was light that enabled Albert Einstein to decipher nature's own formula: $E=mc^2$. Thus it came to pass in 1905 that the world of science finally confirmed that the reconversion, preservation, and conservation of mass energy — yes, electrical energy — fell under the purview of the sun's bountiful light. This law has decreed that energy could not be destroyed or sequestered from existence. No civilization on Earth has failed to pay homage to light in its philosophy, religion, or practical approach to sustaining crops for life itself.

The Refractometer

The science of refraction of light has enabled an examination of plant fluids by a simple field test. The refractometer, a handy tool that is used by grape growers and others, presumes to measure plant sugars, and it does, but the caveats are esoteric and many. Yet they merge with technologies that ask the plant in a manner strangely related to this narrative.

Don Jansen, a former associate of Maynard Murray, M.D., (whose work is the subject of *Fertility From the Ocean Deep*), once objected that his tomatoes grown with the aid of sea water, failed the refraction of light test, even though more sophisticated laboratory results revealed that 50 of the 92 elements in ocean water had made it into the tissue of the tomato tested. The refractometer results were disturbing. Indeed, they were so low that they seemed to cancel out the validity of refracted sunlight to define minerals and identify sugars based on dissolved solids. Jansen reported testeding everything, even 35 percent hydrogen peroxide (H_2O_2). The Brix reading for H_2O_2 astonished this rancher-professor turned agronomist. The 35 percent hydrogen peroxide tested in double digits! What exactly does the sunlight-based refractometer measure? How can it be sugars, since there are none in hydrogen peroxide?

The refractometer measures total dissolved solids. It actually measures the refraction of light, the bending of light rays by the atoms of minerals, ions dissolved in water solution. The scale is actually weight percent sucrose in water. The typical scale is 0 to 32 sucrose. There are four principal factors that affect the reading based on refraction of light:

1. The atoms in the dissolved medium.
2. The atomic weight of the elements involved.
3. The covalent bond between atoms. (This is where the sugar molecule comes in. It has a lot of covalent bonds, which accounts for high refraction.)
4. The temperature of the sample. Temperature greatly affects the reading. An automatic temperature-compensating instrument eliminates this deficit. A very old product out of the refrigerator will present a high refractometer reading. When the temperature of the material is lower than ambient, the validity of the refractometer goes out the window. A recalibration is indicated, or a temperature-compensating instrument has to be used.

Finally, the refractometer will not tell you whether sugar is present — you have to know this from knowledge of the plants. In a real sense, the instrument really measures the covalent factor, an area in which the sugar present in plants really shines. It also happens that H_2O_2 is a covalent powerhouse.

The refractometer is not alone in the field test department. Chandler also has a quick test for chlorophyll and sugars as well as dissolved solids, pH and NPK nutrients. Sap in the plant is best tested in the laboratory because advanced instrumentation yields an absolute set of numbers, even to parts per million of what is in the sap for future growth.

"We can read the physiology of the plant," Chandler explains, "coupled with field conditions. This interaction becomes effective. You can't just have a set of numbers and go for the season. Every week you have to adjust the numbers. Leaf analysis is the autopsy of what that leaf has already taken with a different set of standards."

Leaf/Petiole (Stem) Sampling

For maximizing profit from adequate fertilization, plant samples need to be very specific because nutrient content of leaves vary with the location of the leaf on the stalk. The age of the leaf also makes a big difference.

Always inform the lab of the exact leaves sampled (young leaves test differently than old) so the proper standards of nutrient levels can be used for the interpretations. Any leaf sample is better than a guess, but an accurate sample is a much better guide for fertilization. Select one leaf per plant, of the same age and location for each of the composite sample.

Size of Sample

Take 20 to 50 individual leaves or petioles (stems). The number depends on the size of leaf or stem. A larger sample results in too much volume in the lab and can result in sample segregation. A too-small sample amount results in poor aliquots (measurements of ingredients).

Which Leaves?

Take the most recent fully developed leaf. Take the whole leaf for long-growing plants such as citrus, pecan, shrubs, onions, etc. Pinch the leaf off the stem for the leaf sample. Take the petiole (stem) for short-term plants such as cotton, melons, peppers, soybeans, cabbage, etc. For petiole sampling discard the leaf and save the stem. (Include a few leaves for observation.) For grasses, take the whole plant or several clippings.

Representative Sample

Decide on the test station areas. Mark and map this with a management area identification. Test areas should be confined to similar soil types and conditions and should not exceed 5-10 acres. It is better to be very selective in taking samples than to try to use volume as a representative sample. Results of specific samples from a similar area can be averaged with results from other samples for treatment as a whole. However, averaging samples from all over the field does not tell you the variations, which could show major differences that need separate treatment.

For best results, map areas (stations) that were sampled. Sample the same area each time to compare with previous tests. Avoid variations due to sampling of different areas. Sample same specific areas or plants each time as representing larger areas to be treated. As management is intensified, the entire farm can be placed under testing for each 10 to 40 acres. For the best soil fertility and plant nutrition management program, a history of specific site sampling information is better than an average sampling.

Plant Sample Handling

Wash the samples gently before they wilt to remove any contaminates like dust, sweat, etc. Use a non-phosphate detergent (such as Ivory or Joy liquid dishwashing detergent). Lightly rub the surface of each leaf and rinse at least once in clean water. The last rinse should be with distilled water, if possible. Only rinse petioles — do not crush.

Handle with clean hands and place the sample only on clean surfaces or in paper bags. A simple rinse is better than nothing!

Place in a paper bag so leaves and petioles can dry. Do not enclose in airtight plastic. Punch holes if plastic must be used so they do not mold while in transit to the lab. The first thing the lab must do is dry the leaves. The process of drying can be started as soon as leaves are washed. Do not contaminate.

Dry plant samples can be stored for quite some time without deterioration. Dry slowly (100°F for 8 hours or longer). Use only low heat if any. An air conditioner exhaust, vehicle dashboard, hair dryer, etc. works well.

Identification of Sample

This should include date taken, size/age of plant, growth condition, soil moisture level, insect or disease damage, production and fertilizer history, a copy of any previous soil tests, and any other observations that could influence growth. Include all information with the samples in the shipment to the lab. Interpretation of the lab results and recommendations are much better when all information about the soil and crop is furnished.

Crop-Logging

Using multiple sampling dates is a proven method for the best nutrient management. Sample several times at critical plant requirement periods during the growing season to adjust this crop for better results.

Leaf & Petiole Analysis

We will defer a fuller explanation to later chapters on petiole analysis and laboratory procedures. For now it is enough to note that most growers who use these technologies make an analysis in the fall for the next year's crop. Chandler orders up these tests in the beginning of the season and at various stages of growth. On fruit or nut trees, citrus or pecans, this means three or four times a year. Peaches, four times a year. Apples, much the same, each being coordinated with the physiological stage of growth. While the fruit is being formed, it is being fed through the root. The refractometer tells the grower where he is today, at this hour. The tool is useful. It can direct intervention if there is growing time left.

If you take the petiole, first remove the leaf so that it does not draw the moisture, then test the slow-dried petiole with lab instruments. This permits an accurate quantitative as well as qualitative analysis. By way of contrast, a tissue analysis (quick field test) must be viewed according to the stage at sampling. Is it in a wilting stage, a recovery stage? What is the time of day, the state of sunlight, etc.? All can be useful when interpreted properly.

The refractometer calculates total dissolved solids, which may be translated into sugar content. Here, too, variation is the name of the game, and a petiole test in the laboratory will exhibit an answer that is clearer and more accurate than the refractometer sap test in the field. "Those quick tissue tests are highly variable due to sunlight, temperature, moisture conditions, even wind velocity," Chandler sums up. "This requires experienced professional interpretation for accurate conclusions.

The petiole/leaf test, on the other hand, regardless of when it is taken, yields more of a quantitative than qualitative result, which also requires knowing field conditions for selection of proper standards each week."

A nitrate monitoring program of only NO_3 and PO_4 is very beneficial up to a point, but it is often confused with a crop-logging petiole testing program that includes many other important nutrients with standards adjusted for stages of plant growth and needs.

A leaf analysis, in turn, is also a dry destructive test. It tells the grower what has happened in the past, not what is arriving to govern future growth. When Chandler's "ask the plant" dictum is invoked, the objective is to predict future growth. A correct prediction allows remedial action in the soil and/or in the water — drip, center pivot, or flood — or foliar on the leaf. A reliable soil test of available nutrients is also beneficial.

"The plant drinks and eats every day, not once a season. It has only the capacity nature has provided. But we have been taught to put all of that phosphate on as well as most of our nitrogen. Then we side-dress more nitrogen. But when we use our CO_2 soil test to see what is actually available to that plant, plus the petiole sap test, we find the phosphate in a decreasing availability mode with time." So states Chandler, answering the journalist's question much as the plant answers the farmer's question.

It is a matter of record that only 5-15 percent of applied phosphate ends up in this crop. The number-one culprit in tying up phosphate is calcium. Some parts of the answer are found in the source of phosphate, rock phosphate or calcium phosphate. Calcium phosphate is made available by acidification. Rock phosphate becomes acidified in the soil in a manner decreed by nature to be slowly available. Wisconsin determined that sulfur added to rock phosphate enabled soil microbes to convert sulfur to sulfuric acid, which in turn acidified phosphate to make single super-phosphate. This speeded up the natural biological acidification of phosphorus to the available phosphate form.

As advanced as Chandler is in his study, he is the first to admit that "we are just scratching the surface on asking the plant," so answers can adjust the feeding ration as required, not once a year, but week by week, day by day, as the plant grows and needs change. Drip-irrigation now allows a control only wished for a couple of decades ago. With drip zones it is possible to compare row against row.

Enough Water?

Esper K. Chandler, whose insight, life and times are the subject of this book, enlarges upon that old Biblical injunction to include the first great absolute in crop production: water. Ever pragmatic, ever alert to the day-to-day stages of a plant's life and production, Chandler has added to the usual scrutiny of cations and anions and the usual science of soil laboratories a norm all his own. If "ask the plant" is worthy of legal protection, then surely the copyright belongs to Chandler. "Plant, are you getting enough light? Enough water?" By the time the school-men's answer arrives, the information is probably no longer germane.

Listen to Chandler and do not stop listening. Those who know the man will swear that a thousand conversations will yield, not a thousand but thousands of nuances that govern both plant growth and the bot-tom line. "The only time we seem to fill our dams is during hurricane season. But sometimes the Rio Grande Valley has a problem few agrono-mists envisioned: too much water. The plains in West Texas have more grass. During a recent year, excess water called upon good tilth, deep penetration storage, and internal drainage to prevent water-logging subsoils. "The abundance," he says, "impeded the cotton harvest, knocking it back 25 percent or more. Wasted out in the field, the crop nevertheless made a new record at the gin. Quality was down, for which reason prices trembled at bottom. Late season brought on the start of a drought."

A pioneer in helping to bring highly productive drip-irrigation to both the Rio Grande Valley and other parts of Texas and the South, Chandler was forced to observe that often dry land made just as much yield in terms of bales as irrigated acres. Water is both a curse and a blessing. "The example," says Chandler, "is peanuts. The yield can be superb, yet the crop can go to rot unharvested in the field."

After making allowances for the mix of electrical impulses contained in natural light, Chandler says, "Water is the prime requirement of any plant. The first answer I ask the plant to furnish is: Have you got enough or too much water?" Iron levels in the plants are good indica-tors, but not absolute.

Water is the overriding factor, but then the most efficient use of water comes from having balanced nutrition. Chandler does not make statements without supporting proof ready for recitation or quiet study.

Almost any good laboratory has shelves that are swaybacked with research proving that nitrogen and water will grow vegetative volume, but not quality. Under conditions of balanced nutrition, quality and yield become a given. In vegetable crops, quality means food value to the human consumer and shelf life for shippers and retailers.

Nutrient Balances

Now the field of inquiry moves off into the pantheon of humus, minerals and micronutrients, vitamins, hormones and enzymes. Both nature's fecundity and its demand for balance have challenged agronomists since well before Justus von Liebig. The professors and their controlled experiments have answered with alacrity. Chandler went to war from the cotton fields on the Logan Plantation and came out of it directly back into the fields, went into academia, research and the fertilizer industry, and then back into the field. In the process he found many of the settled absolutes somewhat wanting.

In the fullness of time he recalled the little old lady who talked to her potted plants, perhaps even gifting those plants with a measure of her carbon dioxide. Perhaps a person-to-plant relationship wasn't such a bad idea. Chandler probably didn't figure on smooth-talking the plant or offering veiled advice. But he did want to ask valid questions and hope to catch sign language answers. The first question was: What is the nutritional balance in your bloodstream? And he used quick field tissue-test kits for over two decades to find the answer. Wasn't that what doctors do when patients arrive at the emergency room after being hammered on the highway? Sap is to the plant what blood is to the human being. Vital fluids have to flow in either case.

"It is the same thing with a plant," Chandler instructs. "Too often we take a look, throw on nitrogen and watch that vegetative response paint the crop green. When nitrogen goes up, the uptake of minerals, phosphates, calcium, magnesium, and most essential minerals goes down. As a consequence, quality goes down, plant leaves are weakened, thereby inviting insects and diseases." Asking the plant via plant analysis implies a breakdown between the leaf analysis and a petiole analysis, the last a look at the sap bloodstream for future growth. A leaf analysis is an autopsy of sorts, a readout that tells what the plant has gotten in the past.

The Essential Light

Petiole analysis in fast-growing plants is important because of modern irrigation methods that enable injection of precisely what the plant needs when it needs it. The petiole is the stem that connects the leaf to the main stalk of a plant. That is where the flow of nutrients from the roots through the stalk connects to the leaf. When the delivery system for "Let there be light" — the sun — delivers sunshine to the leaf, miraculous chemical reactions take place via the agency of magnesium, chlorophyll, and the resultant photosynthesis. The sun energy source feeds the plant. The process is a wonderment. Elemental nutrients are converted into organic compounds of proteins and carbohydrates as well as hormones and enzymes, the Krebs cycle.

The inventory of nutrients in the sap of the plant that arrives in the leaf via the petiole influences what the light does by way of transforming organic elements into beneficial compounds to aid plant growth and plant quality.

"Let there be light" may be the most meaningful agronomy statement ever written. Clearly, the sun and its revenue of light keys the photosynthesis to be discussed later. But now it is necessary to drive home the point that plants cannot thrive without a full complement of light. This bland statement asks for an explanation usually neglected by consultants and ignored by laboratory scientists who never answer the question because it is seldom asked.

It matters not whether all the fertilizers (plant foods) are in place, the tilth is perfect, water adequate, and the temperature in compliance. Food crops will not produce in the absence of sunlight.

In 1815, a volcano in Indonesia named Tambora exploded in the greatest eruption in modern history. It killed 70,000 people immediately. The local devastation eliminated entire forests and croplands. Famine and pandemics followed. As far away as Europe and China, weather and climate changed. Approximately 50 to 100 cubic miles of material leapt skyward and the resultant ash circled the globe, producing a perpetual darkness at noon over Europe. Failure of subatomic particles from the sun to reach planet Earth resulted in crop failure, starvation and rampaging epizootics.

Winter came to the northern parts of North America as the year without a summer settled in. New England had killing frosts every month of the year. Midday darkness inspired Lord Byron to write the poem Darkness, and Mary Shelley created her terror novel *Frankenstein, or The Modern Prometheus*. There have been similar eruptions of less magnitude in recent years in Mexico, Central America, the Philippines, and other South Pacific islands resulting in years of atmospheric haze.

In 1926, Walter Russell postulated the existence of nutrients known as sub-atomic particles that arrived on light waves as photons. There are 22 subatomic particles lighter than hydrogen. The number one mineral for vitamin D is scandium, the most abundant mineral on the sun. We don't consume much scandium, nor does plant life, yet scandium is the key to vitamin D construction. Scandium has to be sourced via sunlight. It is the number one mineral in the vitamin D receptor. By way of comparison, iodine is number four.

The amount of subatomic material riding in on wavelengths of light is so small, it does not even register on the instrumentation currently in general use. Of the 22 subatomic particles discussed by Walter Russell, starting with alberton and boston and ending with protium, duterium, and tritium, not all have yet become tied to the genetic code of human beings and animals. Light has the flavor of scandium even though scandium is a real mineral on the periodic chart. The issue with plants is an area of investigation that awaits scientists still unborn. Among the others that Walter Russell named are buzzeon, erneston, penrynium, barnardon, eykaon, jamearnon and other colorful names such as gammanon. The rest of the count is too esoteric for our discussion because so little is known of the role they play as nutrients. Number one is the lightest and number 22 is the heaviest, though all are lighter than hydrogen.

For now it is enough to point out that when radiation hits the DNA, it breaks the hydrogen bond, resulting in breakdown of the DNA. DNA is a complex set of proteins designed to interact with all minerals and sub-atomic particles to accept and utilize the minerals it needs and to reject and dispose of minerals it does not need. The minerals being used constantly are opposite to subatomic particles, the role of which is construction of electromagnetic circuits in life forms, plants included. When subatomic particles are trapped in particle accelerators, the measurable life of captivity is perhaps one-tenth of a second. No art of mankind can concentrate a subatomic particle, so it is futile to do much more than understand them. Tangible minerals below hydrogen on the Mendeleev Table invite the subatomic particles according to ratios. Physicists hardly deal with subatomic particles, but agronomists must. They must respond to "Let there be light."

Cut off the sun and nothing happens. This observation asks us to plug in what we know about chlorophyll and photosynthesis. The photon is a unit of energy. This mysterious photon initiates the production of chlorophyll. When sunlight falters, the photons no longer exist and the subatomic particles are canceled out because chlorophyll is not operative. Now we can reap the intelligence that between light and hydrogen there are 22 subatomic particles which are actually components of light.

The Atomic Energy Commission must have had some inkling about radioactivity and plants. For years, atomic waste was spread over deserts and grassland, in an attempt to solve the problem of disposal via dilution and dispersal. Oklahoma even ratified the use of atomic plant waste as a fertilizer with the name of raffinate. All these experiments were disasters because the materials were not rationed according to the law of photons. My authority for that statement is Harold Wills, a former *Newsweek* reporter, who told me he was never permitted to write the story.

Carbon, hydrogen and oxygen are the main elements pulled in out of the air. In certain polluted environments, plants suffer from debilitating intervention, received or rejected through the stomata. According to species and evolutionary development, nature's decree permits respiration shutdown when temperatures threaten to return too much needed water to the air.

Light & Petiole Analysis

Withal, reminds Chandler, "You must have sunshine finding the leaves of the plant in order to get the maximum carbohydrate production — the sugars, starches and so forth — which is the life blood of the plant." Once that exposure has taken place in the leaf, then the fluids flow back down the petiole and enter the mainstream of the plant. If the leaf has plenty of carbohydrates, it sends a signal to the plant to form a sink capable of storing that carbohydrate product. "This is the fruiting site." Plants produce carbohydrates for reproductive fruit, then feed terminal growth, and, lastly, roots. Roots cut out when they starve, thus ending the plant's growth cycle. Lengyel discovered that the PO_4 level in the sap indicated degree of root activity.

Chandler advises, lectures and teaches. These quoted statements seldom stand-alone. They enlarge themselves as the discussion shifts from fruits, vegetables and legumes to grains and crops such as soybeans, peppers, melons, cotton, and half a hundred others. "These crops are indeterminate. They represent a continuous process as long as the flow through the petiole is maintained. Proper nutrients put fruiting sites into place. Once that fruiting site has been satisfied, the rest of the nutrients go into vegetative valence. After the terminal growth is satisfied, the nutrient flow goes back to the stalk, then to the roots." Once the fruit cycle is satisfied and the vegetative demands are met, the root comes next in the pecking order. The root is the last to get a supply of nutrients, courtesy of the petiole. As the root's food supply fades, so does the uptake of nutrients, with the end result being cut-off of the plant's life, seen in the petiole tests weeks before visual signs occur.

Many crop scientists overlook the importance of the position of the petiole. Chandler is at his most precise when the point has to be memorized the way children were once required to learn their multiplication tables. "In sampling the petiole, you have to take the most recent fully expanded leaf near maturity, one that doesn't take anything out of the sap flow. It should be one that doesn't put anything back either. Only then can you measure what is coming from the root to feed the future growth, which will be seen between seven and 21 days either as visual deficiency symptoms or as adequate growth."

Micronutrients are slow in getting to new growth at the top of the plant. They do not translocate. The older leaf at the bottom of the plant translocates NPK into the sap flow to satisfy the new growth. That part of the corn plant or soybean leaf, the older leaf on any plant, will exhibit the major NPK deficiencies. Cotyledon leaves are the first two leaves to appear to initiate growth. When the multifoliate leaves appear, as in the case of pecans, or the trifoliate, as in the case of soybeans, functions change moving up the plant.

In fact, the plant answers even before you ask. It almost shouts out its communication, telling the grower what is coming from the root to serve future growth. That is why the sap in an older leaf can be misleading. "You won't know whether the nutrients came from the soil or from plant translocation," Chandler summarizes. "Translocation means the older leaf is going to give up and die rather than continue to produce carbohydrates. But the younger leaf at the top of the plant will absorb the petiole-translocated nutrients after the fruit is satisfied."

Ask the plant and it will tell you to make adjustments and interpretations. The translation is simple enough. Interpretation has to proceed on an uncommon good-sense basis guided by observations in the field, meaning all of the growth (environmental) factors affecting that plant.

Ask the right questions and the plant and its environment will tell you whether you have surplus moisture or deficient moisture. Heads-up observation will tell you whether the crop has established the kingdom for the year or allowed weed proliferation to assert an uninvited presence. If exchangeable forms of nitrogen, phosphorus and potassium have failed to satisfy exchange expectations during root-bed preparation, the arrival of insects will tell more than you may wish to know about the government of the kingdom at inspection time. What, indeed is the genetic state of that plant? Is it in a fruiting stage or in the vegetative stage preceding the fruiting stage?

All of the above and more have to become part of the formula if chemical balance is to be evaluated correctly. The flow of the sap never lies, not if it is tapped and evaluated correctly. However, only a reliable natural soil test coupled with a plant analysis can tell you if the nutrients are available in the soil for the plants or if they must be added.

The Plant Pump

Many factors affect the root's ability to take up nutrients. Temperature is never a given. Still, its precise gift (or curse) always affects chemical actions. Too cold, the plant struggles. Too hot, wilt asserts itself. The stomata exist to regulate the respiration of the plant. They govern the amount of air and water taken in. If the plant gets the nutrients it needs from the roots, then it will not continue to pump water while on the hunt for more nutrients. Instead, the stomata close. When they close, water is used more efficiently. If nitrogen and other nutrients are deficient, the plant pump stays on duty until it finds enough food, then it closes down. The quality of nutrient balance will affect the chemistry and physiology of the plant.

Chandler assembled his observations starting with childhood. The old saying about mistreated plants screaming like a stuck hog takes him back to his Louisiana youth. "Cotton was planted right outside a box-type family home which had enough space between the one-board walls to allow wind to set up a whistle coming through. Grandmother would say, 'Hush, kids. If you'll be quiet you can hear that cotton grow with a pleasing sound.' When the plant was in stress, she would say, 'Hush, that plant is cryin' for moisture.'"

The Language of Plants

Listen to Chandler. "When we ask the plant, we have to understand that it talks to us in its own language. It will talk in a chemical form via laboratory analysis. It also gives us enough visual signals to fill a book. I'm talking about deficiency symptoms, wilting in the heat of the day. You can have moist soil and a wilting plant because it can't pick up enough water. We also know that when nutrients are balanced the plant staves off wilt because the stomata close. That plant lets the photosynthesis process take place when the sun shines and you feed the plant or the soil or both properly."

Chandler assembled his fount of knowledge by standing on the shoulders of giants. In each case he added to his inventory of information by exhibiting an interest. Why does the corn plant snap, crackle and pop during a warm summer evening? It literally talks to you, Chandler proposes. He isn't suggesting a metaphor, but a reality. The tools available to the grower are as simple or as complex as the intellectual development of the grower will allow.

It all comes back to tissue analysis versus petiole analysis, this in Chandler's words. Add versus leaf analysis! Tissue analysis is not fully defined in the literature except by Mills and Jones in *Plant Analysis Handbook II,* which has it as "the live tissue tested on site." There are pH and electrical conduction meters, chlorophyll meters, refractometers, and instruments that permit a quick field NPK sap test based on tissue samples. The sap is squeezed out of the tissue in the field. Most scientists agree that this test is not as definitive as destructive plant analysis.

These few notes telegraph lessons in agronomy that will emerge in the chapters that follow. This much is stated here because science is full of traps, cul-de-sacs, false premises, and findings. They separate the so-called conventional from the innovative which refers validation to nature itself.

The Carbon-Water Cycle

The Quest for Carbon Answers

After "Let there be light," there must have been other unrecorded orders: "Let there be chlorophyll," "Let there be carbon," and "Let there be water."

On December 10, 1961, Melvin Calvin addressed the Nobel Prize banquet to tell "Your Majesties, Your Royal Highnesses, Your Excellencies" about his decade-long search for the path of carbon in photosynthesis. It was the same path Chandler would follow.

"No chemical process has a greater importance to the incorporation of atmospheric carbon dioxide into a starch molecule of a green plant under the influence of light from the sun," was the introductory remark by G. Liljestrand of the Academy of Science. "This reaction is the foundation of life, not only for the green plants themselves, but also for all higher animals. This complicated process, the object of intense studies for more than a century has now been unraveled." Calvin unraveled it by detecting the intermediate steps in the reaction.

There is often a small eternity between discovery and recognition of that discovery in the field. Even today, professors with credentials and standing have barely a nodding acquaintance with Melvin Calvin, or the giants on whose shoulders Calvin stood to achieve his Nobel prize-winning success.

Carbon and hydration make carbohydrates possible, relying on the agency of the sun. Carbohydrates in turn feed the soil's microorganisms. Speculation on how carbohydrates are built hark back to the days of Justus von Liebig, and march forward under the tutelage of the sun. The actors in the drama need not detain us, but for the record they were Adolf von Baeyer, Richard Willstatter, Arthur Stall, and in the 20th century, the students and associates of Calvin.

Now we are obliged to quote Melvin Calvin. "Actually, the route by which animal organisms perform the reverse reaction, that is, the combustion of carbohydrates into carbon dioxide and water, turned out to be the first one to be successfully mapped by Otto Meyerhof and Hans Krebs." Yes, Hans Krebs of Krebs cycle fame. The production of oxygen from water and the complicated collaboration of elements to accomplish carbon dioxide reduction can leave the student enthralled. For the agronomist, these revelations lead to the fact that photosynthesis enables transport of essential growth elements through the petiole and presides over the roles of micronutrients, many of which have yet to be understood. Of maximum interest are the compounds which incorporate carbon dioxide. The dominance of sugar and sugar-acid phosphates reveal themselves. None of these things are possible without sunshine. All depend on water and carbon.

The basics say water and food come first. Leaders of countries hardly pause to comprehend what heavy nitrogen continues to do to the environment.

Measuring geological ages has always bedeviled scholars, but ingenuity surfaces to solve the most difficult problems. Measurement of the radioactivity given off by isotopes of certain elements is now settled science. Commonly this means analyzing the carbon in organic substances to determine the carbon 14 relative to carbon 12, the last being the most common isotope. Science now relies on the existence of subatomic particles arriving with sunlight, many of which have now been identified as essential for human and plant health. Cosmic rays arriving during daylight hours react with atmospheric acids. One reaction changes nitrogen to carbon 14 and hydrogen. Carbon 14 hangs onto its radioactivity; its half-life being computed at 5,730 years. Carbon 14 and carbon 12 react with oxygen to form carbon dioxide, some of which is taken up by plants, some of which lingers to help the CO_2 generated by industrial pollution and internal combustion engines cause global warming. It is easy to see why anhydrous ammonia and excess nitrogen of every stripe may be responsible for much more carbon dioxide than the carbon sink can accommodate.

Dead trees and plants do not take in carbon dioxide. That is why carbon 14 wastes away with time. The process is called radioactive decay. For this reason, the ratio of carbon 12 to 14 serves the geologist in dating whatever once lived.

So far, agriculture has been tardy to admit or indifferent to its role in enhancing carbon dioxide release beyond the capacity of plants and other ways to utilize the product.

Photosynthesis Unraveled

We are left with a clear understanding of the photosynthesis riddle so skillfully unraveled by Melvin Calvin and fellow Nobel Prize winners, including Sir Hans Adolf Krebs, the British biochemist.

Finally, Chandler's comprehension of photosynthesis and its complexities has allowed him to enter the world of organiculture to find a storehouse of knowledge — some of it folklore, some light years ahead of the legislated science that special interests have foisted on agriculture.

All that follows starts with the chemical process that anoints green plants with the task of manufacturing food from water and carbon dioxide using sun-energy chloroplasts, which are chlorophyll substances in leaf cells. Nature's alchemy splits water into hydrogen and oxygen. Oxygen is set free to enable life on the planet. Hydrogen finds refuge in a carrier molecule in an instantaneous process. Hydrogen, with its own sidebar construct, converts carbon dioxide into sugar, glucose and starch.

Water and organic matter go together like ham and eggs, which illustrates the thesis in Ed Faulkner's *Plowman's Folly* and invites a new look at conservation-tillage farming, the epitome of which is grass and forage production.

Here, it is enough to point out that those billions of unpaid soil workers draw their energy from rich, decaying plant material. Water adsorption via the agency of organic matter, a carbon dioxide mix, and control of transpiration holds in escrow a greater potential for water control than most of the dams constructed for that purpose.

It has been touted that applying a half inch of well-digested compost to grasslands and lawns cuts water requirements between 25 and 70 percent. An inch of mulch over the root zones of shrubs saves the same amount of water.

The Krebs Cycle

Thus the findings that Melvin Calvin dissected and then put back together now takes shape as the carbon-water cycle, which holds the key to solving the problems of air pollution, water shortages, and soil erosion. The Krebs cycle is also known as the citric acid cycle. It details the last stage in the oxidation of food nutrients to produce energy. The Krebs cycle takes place in the mitochondria of the cell. Coenzyme A is converted to energy by a cyclic of catalyzed reactions.

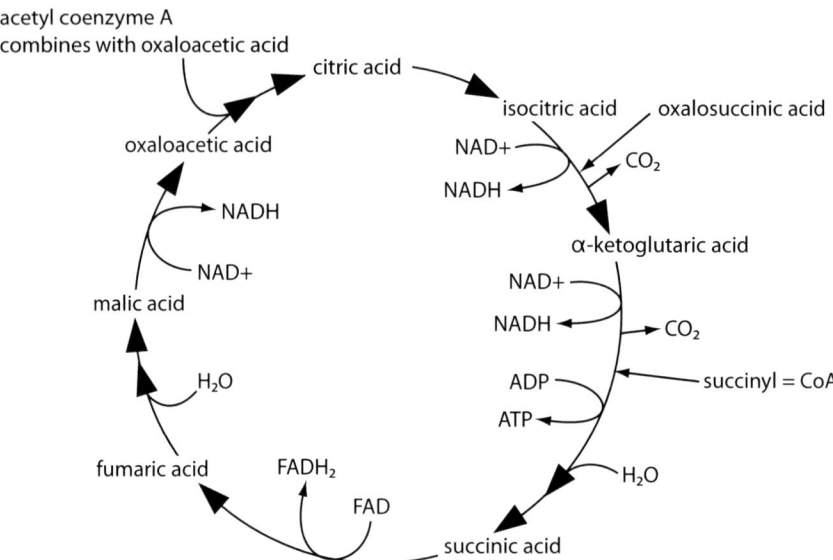

Don't Forget Water

The final footnote is water. The mere presence of organic matter in the soil means an uninterrupted supply to leaves. The stomata entry valve intake mechanism closes when satisfied, and the water rescued from transpiration loss astonishes the mathematician.

An open plant will transpire 99 percent of water taken from the soil under normal conditions, meaning under conditions of low organic matter. When the carbon dioxide levels surrounding the leaf results in closed stomata, the transpiration is cut dramatically. The agronomist can make a calculation, but it is the carbon-water cycle that presides. That excitement rises to new heights about water-use efficiency as Chandler goes about the business of asking the plants.

Fertilizers & Water Use
More Fertilizer = Less Water per Ton

Based on Tests on Irrigated Coastal Bermudagrass in Texas

Elemental Nitrogen applied (Pounds per acre per year)	Fertilizer Cost (Per Year)	Net Return (Per Acre)	Inches of Water Required* (Per 1 Ton of Hay)
0	$7.80	$46.60	12.7
120	24.90	144.10	5.8
240	42.00	220.00	4.0
360	59.10	229.00	3.6
480	76.20	250.60	3.3
600	93.30	265.60	3.0

Prepared by The National Plant Food Institute with data from
Texas A&M University.

* Results for inches of water required are the average of 2 years. Average total
water per year (rainfall and irrigation combined) is 36.52 inches.

The Carbon-Nitrogen Conundrum

Photosynthesis, Carbon, Water & Nitrogen

Lessons that govern photosynthesis and the carbon-water cycle govern the carbon-nitrogen cycle. Over the past 40 years, some scenes have been replayed many times on American farms. A cornbelt farmer walks into his field after a heavy rain. Soil grabs his boots and almost peels them off his feet. There is little carbon left in that soil, and the crop will deliver that message after nitrogen paints the leaves green and cutworms or earworms arrive to stake their claim.

A Minnesota farmer, disgusted with the slow progress his rapeseed is making with a plowpan barrier, resorts to dynamite. His humus has almost vanished, and so have the microbial populations in the soil.

An Illinois farmer, acting on university advice, inserts dry ice into the soil as the plow turns it over, and, sure enough, somewhere a college investigator creates a database proving that this source of carbon "works."

Chandler encountered such reports many times before he departed from the time-honored plains of research and headed for the untried uplands of consultation. People promoting sustainable agriculture assured all those who would listen that with four to five percent organic matter in the soil, there would be enough carbon and enough nitrogen release to make the natural carbon and natural nitrogen cycle work. The problem by 1970 was that very few soils had enough humus to permit classic organic agriculture. This deficit had its genesis in the Tennessee Valley Authority (TVA) fertilizer project of the 1930s, and in the requirements for World War II, which gave the nation the ability to make cheap nitrogen. Since then human beings have doubled the amount of nitrogen in the earth's operating system. The imbalance that resulted stayed hidden as long as there was carbon to buffer the overload. The last half of the 20th century told observant farmers, consultants and schoolmen that cheap nitrogen caused farmers to defy nature. Much of the applied nitrogen left the land.

The Tennessee Valley Authority led unbranded chemical fertilizer research for several decades. Its work became the basis of K's chemical fertilizer knowledge as he participated in several research and educational projects with the TVA.

Synthetic nitrogens from conventional farms and methane contribute to global warming. Some environmentalists suggest that agriculture has turned the Mississippi into an open sewer, one that has toxified a Gulf area the size of any of the smaller New England states. Many find this charge unjust. Cities are the star polluters, and crop agriculture comes in a distant second. Even so, it was the waste of nitrogen that caught and held the attention of Chandler as his laboratory career unfolded. This waste laid bare the fact that carbon and organic matter are necessary to hold nitrogen in the soil profile.

Carbon, Organic Matter & Soils

One of Chandler's associates is Joe Bradford of the Kika de la Garza USDA Research Station in Weslaco, Texas. Ongoing research there faces head-on the challenge of restoring organic matter in soils that have seen organic matter recede to near the vanishing point. Using conservation tillage, the Bradford team's scientifically sound research has been able to rebuild organic matter at the rate of a tenth of a percent per year. Such a performance implemented universally would make a great contribution toward rectifying global atmoshperic change. A tenth of a percent is 1,000 pounds of humus per acre in the top 3 inches of soil, or about 4,000 pounds of raw organic matter of crop residues to be biologically decomposed.

With cheap, factory-made nitrogen available, the real deficit had to be carbon and organic matter, Chandler reasoned. The early history of anhydrous ammonia was one of burning up residual carbon sources (maintained by the original tradition of farming legumes, cover crops and rotations) faster than they could be replaced. Now, the conservation tillage solution has been successful, albeit not universally accepted.

Skeptics shout down the idea exactly the way they shouted down organics 40 years ago.

"Too much of agriculture is dominated by negative ideas," is Chandler's assessment. "When it comes to conservation tillage, we have an evolution of thought. In my early career we called it 'trash farming.'" Chandler recalls his early Dust Bowl experience, where "they came up with the Graham-Hoeme plow, a spring-shank tiller with wide sweeps. When you harvested your wheat crop you would run that across the soil and cut the roots of all vegetation." This was done to stop the use of water by crop residue and weeds. It was the forerunner of conservation tillage. The stubble remained standing. It cut down wind erosion, softened raindrop compaction, and improved water infiltration, thus reducing runoff.

Conservation tillage is not synonymous with no-till. One problem stays on. No one tillage style suits all soil types and all the conditions agriculture has to offer. No-till means what it says, "no till." The usual example is the pasture, not the conservation-till acres on which the glyphosate crutch is widely used.

From Chandler's chair it appears that "we are slow to recognize the fact that organic matter is the source of food for the ecology of the biological populations. They recycle the carbon and the nutrients. These bio-creatures manufacture the nitrogen. The bodies of the microorganisms deliver nitrogen, thereby feeding the cycle."

Justus von Liebig overlooked these salient points when he developed his NPK theory. Possibly it was because he concentrated on minerals, not only NPK, but magnesium and calcium as well. The best assessment nowadays is that von Liebig simply took for granted that the biological contribution of his time would continue. In our own time, the microorganisms are known to make plant foods soluble for root uptake. Missouri's William A. Albrecht was one of the leading proponents of looking beyond NPK, but he dealt first with carbon and nitrogen.

At ground level, the emphasis on soil life seems to be the great contribution that organiculture has made to the soil management debate, but in Chandler's view the entry of bureaucrats into the fray via the Organic Food Production Act has muddied the waters. Soil tests, a prerequisite for any serious look at soils, have been excused by rules and regulations. Organic certification is allowed primarily if "you didn't do this and you didn't do that." Production and sustainability seem to have taken a back seat, according to Chandler.

"It all comes down to one thing: how do you measure?" he says. "What's your ruler? When you take a soil test, the prime requirement is finding the active humus fraction of the raw organic-matter content of the soil, not total carbon. Humus content affects crops every year. Confusion enters when compost is measured for its chemical composition rather than for its life factor and the humus fraction. We always seem to leave an essential part of the puzzle on the back burner or out completely."

This conundrum of nitrogen taking out carbon has prompted almost as much folklore as that of the "Darwinian demon," which allegedly ate 24 hours a day, procreated nightly, and grew to gargantuan dimensions. It wasn't real, but most practitioners consider quite real the practice of salting the field with bituminous coal, Leonardite, and humates — all to return natural carbon to the soil. The use of humic acid in the driplines to shore up the humus shortfall has become so common that it no longer lifts eyebrows for being organic; it is now respectfully conventional.

Chandler came by his insight indirectly. He relates the birth of one idea as follows: "Some 50+ years ago I worked with an entomologist roommate, Ray Young. We were working for Dan Logan, a cotton grower north of Shreveport, Louisiana, the originator of consulting teams scouting cotton insects. Later, Young came up with the practice now termed 'stale bed,' which is still in use. When they harvested the crop, they would pull the bed back up. They would shred the stalks, plow them out, and then build the beds back up. They did that immediately in the fall, and that was the end of the tillage. Now the stalks were atop the bed, which was left over winter. In the spring, the bed was treated with then-conventional herbicides with no more cultivation, thus the name stale bed. This got us away from dealing with wet ground in the spring. If the bed was up, it would dry out faster. That's where they came up with the hip-and-ridger, now the rip-and-ridger, in the Corn Belt. This was a forerunner of conservation tillage. You're not plowing down your residue. You leave it on the surface for the microbes to take it on down."

The anatomy of carbon retention calls upon crop residue to feed the billions of unpaid workers in the soil. Chandler is a pragmatist: first, last and always. How much compost should the grower put on his bed? "Let your pocketbook be your guide," Chandler offers. "The limiting factor is what the soil can take, in most cases." Four inches of a good quality compost is not out of the question, according to Chandler. "Of course, too much compost can deliver too much nitrate too soon. You can also have a lot of potassium, and these can be caustic to germinating seeds. The gardener can water generously, and the soluble salts will move out of the root zone." Moderation served on a regular basis is more nature-friendly than an overdose.

There is a pecking order to the salts. Chloride is highly soluble. Water will take it away. Sodium in the soluble form will melt away in water, albeit much slower than chloride. Potash is as caustic as nitrate — available, but caustic — and slower to leach than nitrate or chloride.

Soil Microorganisms & the Nitrogen Connection

Thus, nature's requirement of balance. To err on the side of generosity is always better than to create a brittle condition for life. Microbes thrive on nitrogen if they have the carbon to balance the nitrate. It's helpful to recall the buffalo herds that used the Great Plains as their pasture. There, nature maintained nitrogen and carbon in equilibrium. Manure droppings delivered enough nitrogen to feed the microbes, and urea from the animals fed the microbes that decomposed carbon. Milling herds ground up the droppings before they moved on.

At the same time, let's not forget microbes that fix atmospheric nitrogen, bacteria such as Azotobacter, Arthrobacter, Achromobacter, Beijerinckia, Clostridium, Nocardia, Nitrosococcus, Nitrosomonas, nitrobacteria, Rhizobium, etc.; plus fungi: Rhodotorula and Pullularia; along with the enzyme nitrogenase and algae: such as Anabaena and Nostoc. They can take nitrogen out of the air if they have the required food source, namely carbons and other minerals to, sustain their lives. That is the heart of the organic sustainable movement: healthy soils with well-balanced minerals and well endowed with humus.

Manure & Microorganisms

Is it possible to have a bona fide organic operation without livestock? Chandler puts it this way: "It is a strong part of my recommendations to have a source of livestock manure. Poultry is included, but the highest order of nutrients is from the dairy industry. You have high protein and a mineral micronutrient balance. A lactating dairy cow also has certain enzymes and hormones in a concentrated form in that manure."

It goes without saying that dairy manure is the source base for many microbial products on the market. Rudolf Steiner required dairy manure for a range of biodynamic preparations, and old man Martin of Medina, Texas, used it with seawater to develop the product now itself called Medina. The soil inoculants that help farmers answer the carbon-nitrogen conundrum are similar to the legume inoculants that came to the fore in the 1930s. If legume inoculants worked, why not soil inoculants? They are, after all, beneficial and naturally occurring soil microorganisms.

"The people who reject soil inoculates because a gallon can't do anything fail to recall that legume inoculant hardly requires ounces per acre." Meat-producing animals drop better carbohydrates than other kinds of livestock.

The next highest species for manure excellence is the egg layer. The balance is superb. After poultry come turkeys and turkey litter. The class of microorganisms changes from dairy to egg layers. If the animal is a dairy goat, the manure is also of the highest order. As manures go, the horse is of a lower order. Horse manure is high in undigested materials.

Nitrogen & Composting

Finally, to achieve the proper carbon-nitrogen balance, composting chores are best accomplished with the static-pile system, according to Chandler. That is not to deny that the windrow system is best suited for bio-solids to eliminate pathogens. Withal, it may be the compost tea that outpaces the spreading of compost. Thousands of Texas acres are currently treated with compost tea. The very idea! Compost tea to solve cotton root rot! Many leading university researchers are now turning to biological solutions because chemicals are failing to control many soil-borne crop pathogens. There are laughs and there are laughs, but the one that rings most merrily is the one that never fails. It is always too early to reject an innovation that pushes the envelope.

Thomas Harr, a Rio Grande Valley composter, puts fish and shrimp waste into compost for still another mix to deal with that carbon-nitrogen conundrum. The microbial action imparts the flavor and the results.

Carbon Dioxide

Chandler's term of choice is precision farming. It exists without the nutrient shortfalls of the greenhouse or the objectives of greenhouse calculus. There are reasons for this. Humus is one. Phosphate uptake is another. Micronutrients and trace minerals are still other reasons. But the one reason that dwarfs the rest is the life in the soil, life denied by the liquid nutrient protocol of the greenhouse.

CO_2 comes off as something of a mystery, if not a bogeyman, nowadays. Some few years ago, Illinois farmers were burying dry ice in the field to get CO_2 release. As a barefoot boy in Louisiana, Chandler walked the cotton rows chopping grass and weeds along with his brother and cousin, standing in the slight shade of the plant to cool their bare feet and to stir the air for a heartier plant. Later, on the Logan plantation, Uncle Morgan would take a mule and drag a log down the rows in the Red River bottom to keep the soil from further cracking and to stir the air and disturb the excesses of CO_2. Aunt Mag would sing to the head-high cotton as she whipped the terminals with a cane pole as she walked telling the plants to quit growing and to set fruit. She stirred the air, too. Nature has decreed the presence of winds at times because nature wants a balance. The plant canopy contains CO_2 and if pressures are not released, stresses in the air will assert themselves.

The working plan for any of the instruments used to determine cation exchange would fill a library shelf. Chandler tells of atomic absorption, spectrometers, inductively coupled plasmas (ICPs), and other costly machines, all with the realization that the most definitive approach is to ask the growing plant. All plant nutrients are salts. As utilized, they are made water-soluble. Salts can be toxic to plants. For this reason, there has to be a balance between CO_2 and sodium, calcium, and magnesium. Albrecht wanted fertilizers to be insoluble, but available. Correct, but after being acted upon by microbes and CO_2, the salts achieve a solubility suitable to the sap of the plant. Balancing involves both the retractable portion and the soluble portion of the nutrient. Nitrates and phosphates in the CO_2 extract are what the plant root takes in. Calcium can tie up phosphate. A high-phosphate soil tested with a strong extracting agent will show high phosphorus in that soil, but the calcium will keep it unavailable to the plant.

The Laboratory Asks the Plant & the Soil

The Laboratory is the Answer

The goal of science is more to foresee than to understand. Those years at the university, in the fields, and on the research plots told Chandler that observation defined the phenomena. To the naked eye the edge of a razor is as straight as an arrow. Under the microscope this straight line can look like the scrawl of a child. In truth, instrumentation available to the researcher permits some fact-founded looks at nature, and yet the instrument hasn't been built that can cope with all of nature's variables.

Chandler made a midlife career change after working nearly 20 years as a farming/research/fertilizer company agronomist and sales executive. He became a farmer and independent crop consultant specializing in soil fertility and plant nutrition, not bug counting or pesticides. At the time he practiced a crop-logging program on cotton petioles across the Cotton Belt with Albin D. Lengyel as tutor.

"I became convinced that the lab was the thing," Chandler reflected, recalling the day he cashed in his savings to enter a partnership with Lengyel to buy the Texas Soil Laboratory from Phoenix Farms, a Groves family enterprise and the largest Rio Grande Valley farm in sugarcane, row crops, vegetables and citrus. Phoenix Farms had acquired the lab from George Schultz, who was still operating it. Chandler and Lengyel inserted "Plant" into the name: Texas Plant & Soil Lab, Inc.

Schultz was a German immigrant who entered the country through Ellis Island. He was credentialed with a Ph.D. yet apparently unemployable. The only job he could find was milking cows in upstate New York. In due time he went to work in a lab testing milk, finally migrating to Wisconsin where a flush of dairies promised full employment for a laboratory worker.

The Florida citrus groves lured the scientist to that sandy clime between 1933 and 1938. In 1938, grove owners in the Rio Grande Valley of Texas — then operating as a citrus group — recruited Schultz to bring his laboratory to deep south Texas. It was a challenge that the sandy soils of Florida could not duplicate, a flood plain dominated by salty, alkaline soils. Wisconsin is beastly cold at times, which is why many of these farmers become snowbirds, Winter Texans, in the lower Rio Grande Valley for a few months of the year. Idle farmers seem to gravitate to sod farming, and some of them got into citrus culture. They handed their problems to Schultz.

It was a poor-boy lab at the time with lots of Rube Goldberg-style apparatuses. He built his own instruments to evaluate in depth the soil's physical conditions, using production studies. He saw the benefits of soluble calcium, actually gypsum. He was one of the early advocates for using sulfur on the soil, because the soil microbes rapidly converted sulfur to sulfuric acid, this to react with free calcium carbonate to form gypsum or calcium sulphate, which improved soil structure so that soluble salts could be leached.

Chandler first crossed Schultz's path in the 1950s. He already had a reputation that brought people from all over the U.S. to his McAllen, Texas door. Usually they had questions about what sulfur would do to alkaline, salty soils. Later, when Chandler took over the lab in 1980 and was attempting to calibrate nutrients to the petiole test, the art that Schultz had used was all but extinct. Back then, the University of Arizona still used a carbon dioxide method for extraction, but the art was being eclipsed by a mandatory modernization and increase in testing speed that complied with the economics of bookkeepers. The laboratory was not the only package transferred to Chandler.

Yes, the purpose of science is to foresee. The method is statistical. The bottom line is the database. If the petiole has something to say, it must do so statistically so that cause can be linked to effect, assuring that no new variables have been introduced. Establishing laws and relationships between facts is the orbit of science, though it is essential to remember that the findings can be warped by the receiving instrument, namely mankind. If there are weaknesses in reasoning, well-designed experiments and procedures will smoke them out. The role of the scientific laboratory is to describe facts, objects, and phenomena in minute detail, so predictions can be made for this plant, this crop, and this season's bottom line. The laboratory has to ask the plant, but it also has to comprehend the answer.

Long before baseball detailed stats to the n*th* degree, tireless workers cast scientific procedures into a working mechanism, albeit not an absolute one. The design of the apparatus and the situation of the worker, much like the conditions of the environment and the fertilizer used, define the outcome of growing plants, correctly or incorrectly. It may be that much of the soil testing in the United States destroys its validity by procedures in handling samples. That was Chandler's assessment when he left the plains of college research and fertilizer education to enter the uplands of practical field and lab agronomy. "We want to be sure we have a representative sample when it hits the laboratory" — and here he trails off to explain how farmer independence is hostile to the procedures a testing laboratory desires. The directions go out, but compliance is spotty, if it exists at all. There seems to be a universal misunderstanding of what is meant by "root zone." The roots feed primarily in the top 24 inches of soil. In most cases the mass of roots rests in the top one foot of soil. Yet the trade has been taught to sample only at 0-6 inches. It does not matter which crop it is, Chandler insists. Properly developed roots will go much deeper if tilth makes it possible because the top 2-4 inches of soil dry quite rapidly, thus affecting availability of moisture and nutrients.

Soil Samples & Testing Soils

"We sample 6 inches, yet when the root establishes its major function, half or more of the total area will be dry." The root then has to feed deeper, therefore soil construction goes much deeper than the top 6 inches. Farming practices create hardpans, preventing deep roots. In certain crops such as turf for athletic fields, the top 3 or 4 inches represent the critical area. As a consequence, the recent industrial fertilize will be in the top 3 or 4 inches.

No amount of laboratory finesse can repair bad sampling. "We want the sample to represent the problem," Chandler instructs. "In the case of turf, we counsel a shallower sample than in a row crop situation. In undisturbed soil, conservation tillage for instance, the top 6 inches are important. That's where most of the fertilizer nutrients are. If the sample goes no deeper than 6 inches, we elude subsoil effects entirely. Too few laboratories concern themselves with the next 12 inches, the

Soil Testing

Soil testing can take the guessing out of soil fertility and plant nutrition. It can be a precise tool or a general ballpark *guess*-timate.

There are three steps to a reliable soil test:
Sampling.
Analytical Processing.
Interpretation and recommendations.

Soil tests can be used as a general representative interpretation or a precise procedure for a specific location.

When? — If a difference can be seen in crop growth (through yield monitors, photos, or observation), soil color, texture, slope and drainage, the soil will vary in test values for each variable. It is never too early or too late, especially when there are problems. Yesterday is always best, but usually soil tests are done at the end of the crop (which is fine) or before planting.

How many? — At least three on very small plots or four (or more) as the size of area sampled is increased.

How deep? — For undisturbed soil such as pastures, hay, no-till, turf or lawns, sports fields, etc. sample 0-6" deep.

For most plants, sample where the majority of the root zone feeding takes place. This is from 0-12" deep.

Consider sampling the subsoil area, as plant roots can reach deeper than the topsoil. Sample at 1' depths as many problems dealing with tilth, hardpans, salts and missing nutrients can be found here.

For trees, shrubs and perennials, sample at a 4' depth or to the soil parent material in 1' increments.

For problems with germinating seeds, sample in the root zone area at 1-3" deep.

Taking the best sample — Get a representative and uniform composite sample of random cores. Combine several cores (be sure to include the exact crust of each core but remove all dry mulch or residue) and thoroughly mix all the soil. Remove all foreign material such as roots, sticks, pebbles, etc.

Sample size — Usually submit a half pint to a full pint in soil volume but some special tests require more.

Container — Use almost any sturdy container to hold the soil. Ziploc sandwich bags are excellent. Do not use thin paper as it deteriorates.

Soil cores (slices) — Cores should be uniform, both sideways and in-depth, from the top to the bottom of hole. Take several randomized uniform cores (slices) across the field /plot of land.

Grids — For small uniform plots, grids are only reliable when used with specialized variable-rate supply equipment.

Sensors — These are still being developed and are only as good as the equipment and operator.

Be precise — Confine the representative/composite samples to areas of known differences. This focuses the treatment by confining correctional treatments to specific areas so general rates can then be used across fields.

Be specific — Samples should be concentrated to a problem area so they can be compared to a nearby normal area.

Be complete — Supply the complete information (goals of the crop to be grown, crop and fertilizer treatment history, any problems, observations, and comments) with each sample. The better the information the better the interpretations and recommendation is to fit your exact soil situation rather than being a general rationalization.

There are three major chemical extraction methods that are used to show the nutrients available to plant roots.

BCSR — Basic Cation Saturation Ratios utilize strong extraction chemicals methods to calculate Cation Exchange Capacity (CEC) by determining % base saturation of the four major cations, Ca, Mg, K, Na and other factors. These cation tests do not calibrate to actual plant uptake (availability).

SLAN — Sufficient Levels Available Nutrients utilize milder chemicals for extraction to determine a rating of sufficiently correlated to crop yields. Actual plant uptake can seldom be calibrated to these ratings.

NLAAN — Natural Level Actual Available Nutrients utilizes a weak natural extraction method using carbon dioxide which occurs naturally and is secreted by plant roots in soils to combine with soil moisture to form weak carbonic acid. These values are rated to actual plant uptake by using plant analysis. CO_2 works well on all soils that grow plants, sandy or clays, acid or alkaline, arid or tropical, etc.

Most labs use strong chemicals for extracting plant nutrients. Only small portions of these large amounts reported are available to plant roots for uptake. This is especially true for potash, magnesium, calcium and sodium. A few other tests are often run, as they are deemed necessary to give a basic fertilizer recommendation, to sell products or to save the user money, and to keep the price low. These include standards of organic matter, the available humus portion, comprehensive salt evaluation (includes total Electrical Conductivity (EC) plus five other soluble minerals), texture, tilth (hard-pans), nitrate, free carbonates, and calcium four ways.

Recommendations and interpretations differ due to extraction procedures and the experience and knowledge of the person or computer formulating them. A good soil lab uses methods and philosophies for rebuilding productive-healthy soils with good organic matter. Emphasis on soil humus content as the key to tilth and nutrient availability for maximum economic yield and good quality. Some labs use the Albrecht system of fertility balancing as a guide to their recommendations.

The goal is to aid growers in getting the most from their investments in crop inputs.

subsoil — that second increment of subsoil tells the farmer whether he's mining the subsoil or bettering the subsoil for long-term, sustainable benefits."

The conventional row crop calls for 0-12" and 12-24" for annual probe penetration and soil testing. The condition of the soil dictates the plane of observation, and the plane of observation dictates the phenomena. This suggests that the lay of the land, the inventory of trouble spots, the weeds, the signs of color, texture, and symptoms evident to the sampler all have to guide the fertilization and depth of the soil, the crop to be grown, and whether to irrigate — an even greater consideration. Chandler is adamant about following these directions. "Especially if we're going to spoon-feed water two or three times a week, meaning we can spoon-feed nutrition, then we need to sample a much smaller area than we would, say, to step up to the next level for center pivot irrigation and nutrient injection," he says. "If we go to flood irrigation, we require a still larger sampling area of 40-80 acres. With dryland cropping, we need to look at 80-120 acres per sample depending on the lay of the land which determines management areas." Management zones and crop value determine the size of sampling and management area.

The soil sample adjusted to the situation is near holy writ in Chandler's viewpoint. It brings up the question of a large area composite representative sampling versus specific precision sampling. The old

Healthy corn roots feed deep, as evident by these two-foot long roots.

grid-sampling concept comes to mind. Chandler has worked with USDA on an intensive sampling program in which a 40-acre field requires a probe every acre using the grid approach. "It was found that the subsoil was more important than the topsoil. Secondly, it is not a matter of grid sampling. It is a matter of characteristic sampling, and this relates to precision agriculture using yield mapping, aerial photography, and infrared images of soils and crops," Chandler explains.

In sales work, the term "hit the bricks" describes a person who abandons the telephone and a plush-bottom office chair to find out what is going on in their territory. The lab worker who walks the soil is in a similar situation as the seller who samples the real world. Precision farming has its own code of requirements. There are services available that monitor the crop several times a year to discern the different stresses revealed by the crops. This requires ground-truthing by consultants, thus enabling specific sampling. While photos can reveal differences in the crop, causes must be documented on the ground. Most soils are quite diverse across an area, therefore precision requires samples representing discernable differences. Take a 100-acre field with several sampling sites. An inventory of top- and subsoils makes a statement. Can you average those differences and treat the whole field? Or must those acres be brought to a norm with calcium or phosphate or potash or magnesium, etc., finally then treating the whole field the same?

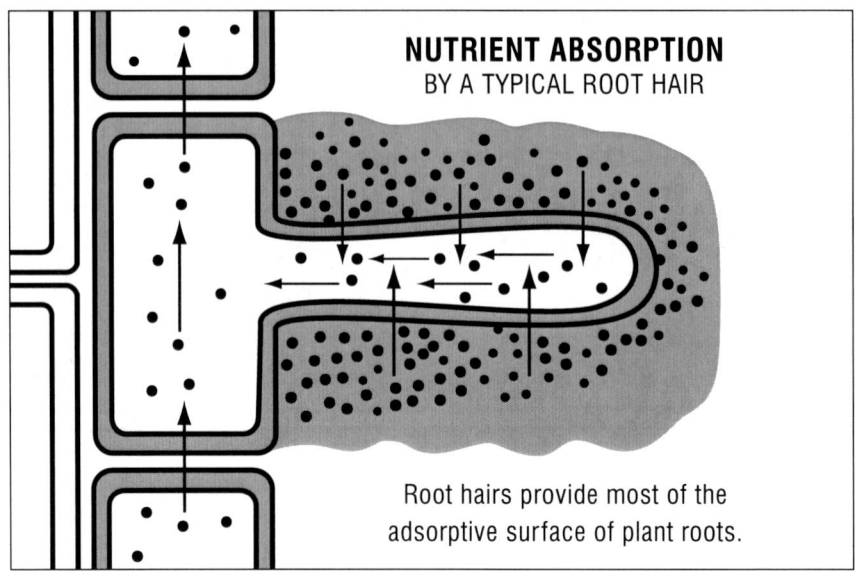

NUTRIENT ABSORPTION
BY A TYPICAL ROOT HAIR

Root hairs provide most of the
adsorptive surface of plant roots.

Young root hairs are the most active feeders.

Sampling is so variable that it always calls into question the high science expected from a laboratory. Chandler will discuss sampling at the drop of a hat, and if time permits, the afternoon can become an evening before he leaves the subject. "You can average Pike's Peak and the floor of the desert, and on the average the sample appears valid. Likewise, standing with one foot in boiling water and one foot on a block of ice, on the average you should be comfortable. The expanses of a field averaged — all other aspects of sampling being kept in mind — enable spot treatment of the worst parts, a holdback on the above-average parts, and overall treatment as the final blessing for the crop."

For efficiency, the sample must be labeled and dried. When the sample arrives, regardless of shape or package, a representative portion of it must be sequestered. Too many labs just take in the sample, stick it in an oven so all samples receive uniform heat, and then proceed to grind using only the portion that falls through first, thus segregating a sample. "Number one, we do not put heat on our samples," Chandler tells visitors. "Warning! When you put heat on a sample, you change the physical and chemical makeup of that sample. The two main defects from a heated sample are a change in pH and a change in cation exchange. The objective is to test that soil under conditions as near as possible to those in the field. You need it air dry, but not completely dry.

Then we prepare the sample, air-drying having been accomplished not in an air conditioned office, but in a room with the tropical air a fan can deliver. Sometimes the sun is used together with the air moved by fans. We want a high percentage of soil naturally dried!"

The Chandler-Lengyel system avoids the pitfalls of, say, college labs with student help and samples mixed without attention to the size of the field sampled, its mix, and the consequences of grinding. Dismissing clods that do not grind easily is tantamount to dismissing part of the sample. If clods are not reduced to a pulverized mass, the sample becomes anemic and still another variable is installed into the effort. Chandler insists that that is what happens to a lot of soil testing. He makes a point of having a total grind, not allowing workers to screen out what for the moment is inconvenient.

A harsh metallic grinder will create variables not accounted for by the procedure. A gentle flail grinder with plastic flails avoids the heat of friction and a change in the chemistry of the sample. This concern for the sample's integrity enables Chandler to call the test the "natural soil test method." "We're mimicking the plant root in extracting nutrients out of the field conditions that the sample represents. Therefore we keep the sample as natural as possible. When you measure your preparations and go through your procedures, then you are as near as possible to what the root is extracting," he explains.

The root can't go to the drug store and buy an extracting solution. The lab can. Therefore it is incumbent on the laboratory to purchase wisely, keeping the "natural' objective in mind. Chandler uses distilled water and purified soda fountain food-grade carbon dioxide. No frozen or volitive chemicals are introduced during the procedure. The CO_2 is bubbled through a special sparger as established by university procedures. Parenthetically, it may be noted that the procedure Chandler uses can be found on the Internet and in textbooks that explain the carbon dioxide extraction method. One signal point must be kept in mind. Whether a plant is grown in desert soil or in a tropical soil or in the soil of a temperate zone, the root performs its extraction via the same method.

Chandler marvels at the efficacy of the CO_2 method. All soil samples come up with a set of values that the plant can actually respond to. "We don't have to go through the cation exchange and a percent-base saturation to calculate the capability of roots to uptake the most important elements," Chandler explains. The leader of all nutrients is calcium, exactly as explained by Albrecht and his era of agronomists. Thus when the CO_2 value gets above a certain level — about 1,000 ppm — the calcium in that soil will be more than adequate for several decades of farming, assuming that abuse is avoided. Under the CO_2 method, below 1,000 ppm and approaching the 500 ppm level, the situation makes possible soil mining, and a race to the production floor to follow. Calcium levels are often found below 250 ppm. Plant feeding can be calibrated to provide the required calcium according to this method of analysis.

Historically, the CO_2 method was developed as a procedure for cation exchange capacity audit of alkaline, saline, or sodic soils, soils short of soluble calcium in spite of their high pH readings. Most of the labs west of New Mexico used the CO_2 method once upon a time. Nowadays it's been abandoned as too cumbersome. Cumbersome or not, abandoning has meant the finer points of the scientific system, because the plane of observation has been obscured.

Nature is difficult to automate. Old "Professor" Schultz was a master of homemade automation, and Chandler followed in his footsteps. Possibly both men realized that the real master of Rube Goldberg automation is nature herself.

Basically, you can set up no more than six samples in a row with the CO_2 extraction procedure. If you go over six, you cannot properly acidify the soil. "If we split hairs," Chandler adds, "we probably could do only about three or four samples at a time." The soil makes the hair-splitting difference. If the soil is alkaline or salty, the prospects of getting a valid readout are different than the same procedure with a deep sandy soil as in coastal plains states or alluvial soils in the Midwest. In truth, Chandler has not found it necessary to split hairs at the current levels of crop management.

The anatomy of the soil must be understood in terms of its own dynamic contribution, independent of the CO_2 that the roots deliver to it. Roots secrete carbon dioxide. This mixes with the soil moisture to form a weak carbonic acid. Over the years, investigators have determined that the CO_2 testing medium should be equal to what a root delivers in about six months. That is the standard that Chandler prefers. These values have been calibrated using plant analysis according to what the plant actually picks up.

Again, the plane of observation has furnished the phenomenon in the case of potash. Chandler maintains that cation exchange capacity (CEC) readouts suggest availability that simply isn't there. "When I started carrying a quick tissue-test field kit and then went to the petiole test, I couldn't calibrate anyone's method, even after using dozens of different methods. Not one test could tell me how much potash, calcium, and magnesium was actually available to plants in fact rather than in theory," he explains. "Add phosphate to that, too."

The first man Chandler worked with on the petiole test was Lengyel. The CO_2 test won out for scientific reliability. In 1980, the year Chandler took over the lab, CO_2 extraction was still official in Arizona. It was during that era that Chandler abandoned the Texas A&M extraction method in favor of the CO_2 method. There were lots of exchanges in those days between Texas Plant & Soil Lab and A&M, with the greater reliability of the CO_2 method bumping the CEC method as far as Chandler was concerned. During Chandler's first year at the laboratory operation, "we ran both methods: Texas A&M University and CO_2 extraction on all soil tests with occasional check samples to other labs. It was the petiole tests that revealed a serious disconnect between what the CEC readout said and what the plant was actually getting. The principles involved became obvious once they had been stated. But what really put CO_2 extraction over the top was the arrival of drip-irrigation. Here you had zones and side-by-side rows where you could put in one combination of plant foods vs. another, and in a week you could start picking up the difference. In the case of phosphate, this was dramatic. We could go from a low phosphate uptake to a high uptake in a week or less." Unfortunately, there was not time or money to appropriately document the massive field work.

The numbers that came to make up the working database told the story. Phosphate uptake told of the health of the root because phosphorus is picked up by young root hairs on the growing tip of the plant. The plant eats every day. This realization clashed with the conventional wisdom that called for a season's worth of phosphate to be applied before the crop went in. That advice failed to service multi-fruiting crops such as soybeans, melons, cotton, peppers and dozens more. When the peak fruiting period arrives, phosphorus starts running out, the crop stops getting enough phosphate, and roots cut out to slow terminal and fruiting growth before genetic potential is reached.

The petiole told Chandler that he could spoon-feed phosphate by running it in the drip-irrigation line. The drip from a pencil-thin hose was simply stretched out down the row at first, but now a technology has developed for burying the hoses at various depths for various crops. The drip hoses last for multiple years with technology rapidly improving. Then the conservation-tillage connection is self-evident. When used, this system leaves the hoses undisturbed year after year, and soil humus is rebuilt faster without tillage.

Starter Phosphorus

Profitable crops depend upon uniform stands with early root systems for fast growth, early fruiting and maturity. Maximum Economic Yields means lowest unit cost of production for producing profits in a competitive marketplace. This requires using all Best Management Practices. Good seed quality of adaptive varieties on good mature seed beds is the first step.

Phosphorus (P) is the foundation for good roots, early fruiting and maturity. After many years of observing the benefits of adequate phosphate on crops, there is no doubt that uniformity of maturity for early harvest results in improved yields and efficiency. Balanced fertility is essential, but phosphorus early is the foundation for profits.

Starter or pop-up fertilizer benefits is a source of debate for many agronomists. Most vegetable growers see the benefit of early uniform harvest and are most likely to over fertilize. There are hazards to using starter fertilizer too near the seed and this scares off many advisors from recommending it. It is easy to accentuate the negative with "Just don't do it." In the battle for profitability accentuating the positive can mean survival.

Placement of P for maximum benefit has been shown by research to be most beneficial when the germinating seed's first root immediately hits a band of high phosphate — especially in cold soils. It is safer to place the band of starter fertilizer 2 to 4 inches to the side and below the seed, but the benefit of early phosphate is reduced by each $1/2$ inch of distance away from the seed. Directly below and adjacent to the seed can be the most beneficial. Dr. F. L. Fisher of Texas A&M University was an early source of research with starter phosphorus.

Phosphorus by itself is seldom harmful. It is the other plant nutrients in fertilizer that have the salt effect that can harm germination and plant growth. Nitrogen, potash and sulfur have high salt indexes and should be kept to a minimum even though some can be beneficial early as can micronutrients of zinc, iron, manganese, copper and boron in minute amounts. Phosphorus can be beneficially sprayed directly on the seed but caution is advised. Too much of a good thing can be harmful. When possible, it is best to apply starter fertilizer directly to the soil below the seed or to the covering soil before it hits the seed. *Do not apply starter fertilizer directly on the seed.* Spraying starter on the soil behind the packer wheel has helped. Use a high-purity fertilizer such as a foliar formula to reduce the salt index.

Only a quart per acre of a high-phosphorus formula next to the seed can produce benefits. It is generally recommended that 1 to 3 gallons per acre of a high-phosphorus formula be used that contains a balance of nutrients. This should be mixed with at least enough water to allow accurate application, 5 gal/acre total is essential with 10 gallons much better for dilution as well as accuracy of application.

Additives such as humic acid or polymers such as Amisorb or other adjurants including plant growth hormones and soil innoculants can be beneficial in increasing nutrient uptake.

Too much nitrogen (N) too early on fruiting crops delays fruiting and increases costs. Dr. Charles Stichler and Dr. Mark McFarland of Texas Agricultural Extension Serivce have a publication *Crop Nutrient Needs* that illustrates when different nutrients are needed in a crop. Cotton especially is harmed by too much nitrogen too early, less than 15% is needed before fruiting. Only a small amount of phosphorus is needed early, but it is essential to have a balance of needed nutrients.

How to know what is needed and when? Start with a soil test that predicts what is available. Then test the plants to see if they are getting what is needed at each stage of growth. A sap (petiole) test of the whole plant at the 4- to 6-inch stage can show what soluble nutrients are in the sap for future growth. Correct with foliar, soil or water applied nutrients early to improve performance. For best results select an advisor that may be an independent consultant, a farm supply advisor or extension agent and a lab that knows the proper standards to apply.

Mepiquat chloride is a slimy product that stops whole plants and roots in their tracks, so the plant has to set fruit rather than grow vegetation. Here again the petiole test will tell whether the root has stopped because the phosphate drops when the roots stop. The phosphate will drop several weeks before top growth cuts off its normal growth.

Petiole Testing

The plant root feeds and drinks every day, just like a growing youngster. This observation calls into question the practice of feeding the plant once a year. With the expectation of a maximum yield a few months later, by checking the plant daily or weekly — crop-logging — the flow of its bloodstream, so to speak, telegraphs its future growth for 7-21 days. When a suitable inventory in the soil fails for any reason, the sap test enables a saving action on time and in compliance with production objectives. "I can't stress it too much," Chandler always reminds us. "The leaf test defines the past. It is valuable, but not really as valuable as petiole testing for the future."

Cation Exchange Capacity

This is the subtle way in which some of organiculture's most vital ideas are sidling into the agriculture that once spat out the word "organic" as something tasteless and unworthy of high science. The syllogism completes itself. The higher the humus, the better the cation exchanges. Cation exchange is a function of the soil's texture. The more clay, the higher the cation exchange. Put humus into the mix, and the cation exchange goes up, hence more available nutrients. The two have to be bifurcated for total analysis. In truth, it was the presence of humus and organic matter that made the salt fertilizers perform so well in the early days, but those salt fertilizers along with rescue chemistry burned away that asset of native soil humus. Today, we are "attempting to recapture the values," Chandler says.

Precision Plant
Nutrient Monitoring Program
"Ask the Plant"

The Importance of Monitoring Plant Nutrition

Plant nutrient monitoring is the foundation of any plant nutrition program. Unfortunately, it is often the most overlooked and deficient part of many growers' management program. Monitoring plant nutrition is an important attribute of any good plant-monitoring program for more efficient agricultural production. Plant analysis is a very precise procedure as to sampling, handling and weighing aliquots. Only $\frac{1}{2}$ of a gram is analyzed using wet digestion methods in the lab. These methods have proven more reliable than dry combustion.

What is Plant Nutrition Monitoring?

This is a very simple question to answer. It is keeping track of the nutritional status in the plant at certain stages. There are four major terms that are important in understanding a very technical and efficient program.

Crop-logging is keeping a record of nutrition at all stages of growth where requirements vary with age.

Tissue Testing comprises quick tests done in the field such as nutrients (mostly on sap — NO_3, PO_4 only is called nitrate monitoring), Brix (sugars), chlorophyll, pH, EC (salts), etc. and are highly variable due to environmental conditions at the time, moisture stress, light, temperature, etc. These require professional experience to fully understand their significance.

Plant (leaf) Analysis Tests (% N and P + minerals and micronutrients) are done on slower growing plants that show the nutrient uptake in a mature leaf, that has occurred in the past and is referred to as autopsy or post mortem, very useful with trees.

Petiole (sap) Testing is done on the sap in the most recent fully developed petiole (stem) of a leaf on a fast-growing plant that has a high sap flow. It shows the nutrients that are available for future growth that will be seen in the plant in 7 to 21 days. When done properly, it is a precise program where standards are adjusted according to the plant's stage of growth, fruiting potential and field conditions.

Unlike pesticide management specialists, where the first thing that comes to mind when the word "monitor" is heard is keeping tabs on pest populations, plant nutrition monitoring also involves keeping track of soil moisture, soil nutrient status, age and development of plants, plant nutrition (via petiole or leaf samples), weather monitoring, measuring transpiration (ETs), plant growth, genetic physiology requirements at each stage of growth, etc. Any test beats a guess, but with precision sampling, knowledgeable lab tests and experience of which standards to be used greatly affect the return on investment — cheap tests save on harvest costs!

When discussing leaf/petiole monitoring, the analogy could be made with the instrument panel on a pickup truck. All the gauges on the panel keep us informed as to what is happening with the truck's various systems, which are complicated and getting more so all the time. Without this instrument panel, we would know almost nothing about the truck except when the engine was running, when it was moving, and when it ran out of gas, overheated, or the battery went dead. The same can be said of leaf/petiole monitoring. It keeps us informed about the operation of our crop, which is an extremely complex biological system. With only visual monitoring (missing the instrument panel) we are almost completely in the dark as to how the crop will operate in the future and as a result, we are not able to make the best decision for obtaining the maximum profit potential.

Why is Petiole/Leaf Monitoring Important?

This is also an obvious question that needs to be answered. Monitoring is asking the plants what nutrients, how much and when they are needed to grow crops better. Another reason for monitoring is to stop applying too much or too little fertilizer (ask the plant what it needs). One of the underlying goals of sustainable crop production is to minimize crop inputs until the crop shows the need, which increases production potential. Monitoring plant nutrients is the way this is achieved, since we should only add inputs when monitoring tells us it is necessary for increased yields and quality.

Some Important Attributes
of a Monitoring Program

Reliability — This may seem obvious, but shouldn't be taken for granted. A monitoring program needs to be reliable for punctuality (regular schedule) and precise as to the part of plant sampled. Sample the most recent fully developed leaf, not older or younger, each time. Take it at the same stage of the plant's physiological development (nutrient requirements change with age), and note field conditions for stresses that help or hinder development. Anyone who follows the procedures for the program will obtain similar results. Developing reliable methods for estimating nutrient status depends upon selecting a representative portion of the majority of the field. Quantification of a monitoring scheme helps a great deal in increasing the reliability of a monitoring program. For example, instead of reporting visual symptoms such as chlorotic leaves are light, moderate or heavy, after the damage, it is more beneficial to measure the levels of micronutrients in the petiole test to correct problems, before visual symptoms are evident.

Consistency — It is important that a monitoring program is carried out in the same manner following the same sampling procedure (area, plant part and age) each time it is done so that the results are consistent from one monitoring to the next. For example, when measuring soil moisture don't take the sample from one area one time and then other areas of the field the next time. The same is true for taking the proper petioles of plants. The most recent fully expanded leaf or stem is used, and that changes with the stage of growth. It is more accurate to take samples from the same area, the same number each time, ensuring that the area sampled is representative of the overall field so actual changes in the plant are recorded, not varying soil characteristics.

Frequency — Monitoring should be done frequently enough to pick up trends or to spot developing problems. "Frequently enough" means that you can watch the problem develop and still allow yourself enough lead time to consider management options. One of the best examples of this is monitoring nitrogen levels. Petiole sap tests show what you should observe visually in 7-21 days as symptoms such as "nodes above white flowers," nutrient deficiencies, or stunting, mean planning ahead. For example, if you suspect a decline in blooms, what you would do about this potential problem would probably differ considerably if, for instance, you couldn't get back to check the field for at least a week, as opposed to a situation in which you knew you would be back in a week to apply more nitrogen based on the lab results. Irrigation systems and programmed schedules now are the most efficient method of feeding plants by injecting nutrients through the water. It is very accurate and efficient, when done properly with drip, sprinkler or flood. Foliar feeding can also be very effective if plant physiology is understood and is best applied in late afternoon or evening when the plant is inhaling.

Speed — When growers and consultants are asked the question, "What, besides money, do we never have enough of?" Invariably, the answer is "time." For a monitoring program to be useful to growers or consultants it must be timely and on a precise schedule for recording plant progress. However, and this is apparent, there is an inverse relationship between time spent monitoring and reliability. The more time spent monitoring, the more reliable the results will be. Therefore, the more efficient the monitoring is, the more cost effective the program is overall. Field sampling and lab analysis costs are only a small fraction of the overall crop production expense — it has the highest return on investment of any crop input.

Written Records — Experience shows this is most often the biggest weakness in growers' or consultants' monitoring programs, particularly in field observations. Very few people keep written records of their monitoring program. If you think otherwise, ask your fellow growers or consultants if they keep written records of their nutrient monitoring program. If you need further proof of this weakness in monitoring programs, try to find computer data management software designed for farming operations that contains a decent component for recording and summarizing information on nutrient monitoring programs. On the Internet, there are software packages that have over 40 different components for handling every imaginable type of data, yet not one of them handled nutrient management monitoring. Although about ten of these modules help growers keep track of various types of data related to spraying for pests, such as spray recommendations, pesticide use reports, and workers safety. Amazingly there was no option for recording nutrient management data that growers or consultants can use to make good fertilizer application decisions! It is apparent that the lack of such components is a good indication of the low priority nutrient management monitoring occupies on the priority lists of many people in agriculture.

Plants Eat & Drink Every Day Just Like Animals!

Why are written records of nutrient management programs so important? All plants start with their roots in the soil looking for nutrients and water. It is through calibrating the nutrient levels and crop performances that allows constant improvement in the knowledge of economic thresholds, which is how best to minimize control procedures, whether they are chemical sprays or anything else for better profits. Control procedures should be justified, chemical or otherwise. The only way to accurately analyze the effectiveness of nutrient management programs is to look at written records of the monitoring test data and field (environmental) conditions at each stage of development, weekly for petioles. No one would consider trying to manage the financial aspects of their farming operation without keeping accurate financial records. Why should nutrient management be any different?

It is hoped that a strong case for the role of plant nutrient monitoring in crops has been made. Minimizing crop inputs is best done by knowing as much as possible about how a crop is functioning. Let's keep our crop's "instrument panel" up to date with the latest monitoring "gauges" so that we don't run out of gas unexpectedly.

Procedures

A successful Plant Nutrition Monitoring system includes precise sampling, knowledgeable lab testing and experienced recommendations.

Planning — This is one of the essential ingredients in a successful (profitable) program. Growers with the desire to make the best better use a team of knowledgeable labs with experienced recommendations, qualified advisors/consultants fieldpersons and reliable suppliers.

It is absolutely essential to minimize variations in sampling, handling and testing procedures. Map and ID fields to be in crop-logging program, mark and ID location of each representative test site, even GPS coordinates.

Records — Document as much as possible, such as: owner's name, contact info e-mail, address, phone, fax, cell phone, physical location of office or shop. Crop, variety, date of planting (or age of first sample). Plant spacing, row width, etc. Fertilizer, crop yield history and goals. Type of irrigation or dryland.

Taking samples — The area of the field selected should be representative of the majority of the crop grown on that field and avoid bad or very good areas. Mark a spot and start sampling at that exact spot every time. Move to a new area if later growth shows the initial choice was poor (and note the change on the field information sheet).

The starting point should be over 50' from the end of rows or side of a field. Sample one plant every third or fifth row (to avoid cultural patterns) at a diagonal across rows incorporating the majority of the field. Large numbers of samples (like melons) can be taken from every row.

Take the most recent fully expanded leaf (petiole), as these petioles do not take nutrients from the sap flowing from the roots to the leaf for the photosynthesis processes. Nutrients, especially carbohydrates in the sap, flow to fruiting sites first, then to vegetative terminal growth and finally, if any is left over, to the roots to keep the plants producing by avoiding cut-out. Root activity, indicated by PO_4 in the sap, is essential to high yields and quality for multi-fruiting plants. Each crop requires precision samples, which are absolutely essential to prevent variation. Young leaves take up nutrients in the sap to grow — thus sap tests lower. Older leaves give up NO_3 and PO_4 to feed new growth — thus sap tests higher.

Record field growing condition notes on the plant information sheets as each field is sampled for accuracy. Proper field notes, sampling, cleaning, packaging and shipping is absolutely essential.

Petiole & Leaf Samples

How many? Plants are about 95% water, so a good sample should be 2 grams dry (a good hand-full when fresh).

How often? Every seven days for rapidly developing plants or 10-14 days for slower plants. This is essential since needs change rapidly as plants progress from vegetative to reproductive to maturing stages, so standards must be adjusted for best returns on investments.

Multi-Fruiting Crops — Use the petioles and take the entire leaf stem from stalk to leaf, discarding the leaves (send 20-30 to test).

Cotton — Count from the terminal (greater than1"), to about the fourth leaf (the first fully expanded petiole/stem leaf). Collect 20 if greater than 4" or 30 if less than 3". Include three separated leaves to help choose proper standards. Sample weekly.

Soybeans — Same as cotton. Discard the three trifoliate leaves and include two complete typical ones for observations. Sample weekly.

Cucurbits (melons, cucumbers, cantaloupes, etc.) — Count from the growing tip of a main runner to the first fully expanded, mature leaf (about number six or seven). Take the full stem/petiole. Collect 40 if they are less than 2" (length of a thumb), 30 if 3" or 25 if 4". Include three separated leaves. Sample weekly.

Peppers — Use the first fully expanded/mature leaf. For stems less than 1" send 40 entire petioles (and a leaf). For 2" size send 35 petioles (plus three leaves). For stems greater than 3" send 25 and three leaves weekly for observation. Sample weekly.

Tomatoes — Use about the third or fourth most recent, fully expanded leaf and stem at a fruiting bloom. Collect 8-12 weekly.

Potatoes — Use the most recent fully extended leaf stem. Collect 10-15 weekly.

Brassicas (cabbage, cauliflower, broccoli) and lettuce — Take samples from the center of the whorl outward to the first fully expanded leaf or full stem and leaf. Separate the bottom 6" of mid-rib and strip off the leaf portion. Collect 6-12 depending on size.

Celery — Use the first full petiole above the most recently developed main stalk blade. Collect 8-12.

Beans & Peas — Submit six to ten early, young whole plants and petioles or the most recent fully developed leaves. Collect 10-15 samples.

Root Crops (onions, beets, carrots, kohlrabi, etc.) — Collect from the center outward to the most recent fully developed leaf/stem. Collect 10-15 samples.

Legumes (alfalfa, clover, peanuts, etc.) — Collect from the tip down to and including the first fully developed leaf. Collect 8-12 samples.

Sunflowers — Collect from the growing terminal down to the first fully expanded leaf/petiole. Use 8-12 petioles at 7-14 day intervals.

Hay, Forages, Turf & Pasture Grasses — Use whole stems and blades at harvest or clipping height. Collect a heaping handful of clippings for a representative composite sample. Take several composite handfuls for hay that can be grasped by thumb and forefinger. Be sure it is representative of the majority of the field. Large samples can result in segregations.

Fruit & Nut Crops

Pecans — Use the most recent fully developed shoots. Take pairs of leaves from the middle of the leaflet. Use 20 pairs at four critical stages of development.

Fruit Trees (peaches, plums, pears, apples, etc.) — Use the most recent fully developed leaves from the base of the fruiting sites. Collect 12-20 leaves.

Strawberries — Use the most recent fully developed leaf/petiole. Collect 25-40 samples, depending on size.

Blueberries and Raspberries, etc. — Use the youngest fully developed leaves from the primary canes. Collect 30-40 leaves.

Grapes — Test the petioles and leaves nearest the fruit cluster. Use 10-20 separate leaves and petioles.

Citrus — Use the most recent fully developed leaf. Also, for a full evaluation, take the oldest leaf on the same limb (showing nutrient transfer, especially potash, N and P). Mineral and micronutrients do not transfer, so foliar applied micro-nutrients stay on that leaf, while new leaves may be struggling in a deficiency zone. Take leaves from fruiting sites for this year's crop and non-bearing leaves for the next year's crop analysis.

Single-Stage Fruiting Crops

Corn — At 21 days take the entire plant, cut off roots, and rinse thoroughly. Collect 8-12 leaves from the entire plants. Boosting balanced nutrient/adjuvants at this stage can increase rows of grain.

Boot stage — Use the most recent fully exposed dewlap leaf. Cut 8-12 leaves precisely at dewlap. Take the base 6" of the midrib and run a petiole (sap) test to see the flow of nutrient for future growth. The middle 6" of the leaf will show the past history of nutrient status to date. Test the leaf for % N, P and micronutrients.

At Pollination — When at dark silk, sample the ear leaf, as with the boot stage. Nutrients can be used in water or as a foliar spray.

Cereal Grains (wheat, rice, oats, flax, etc.) — Use the most recent fully developed whole stems and leaves. Collect 12-20 leaves at 14-21 day intervals.

Sugarcane — This crop has the most researched and precise plant analysis program published. On a two- to three-week schedule from a marked site, take five to seven samples of one full stalk/row on a five skip-row diagonal pattern. Cut five to seven leaves exactly where the dewlap is fully exposed. Cut the exact middle 6" of the leaves and strip the midrib. Record growth development. Also vital information can be obtained from the entire sheath of these leaves, especially for moisture monitoring, potash and other minerals.

Interpretation Guide to Soil Test Reports

The following is how laboratory results are reported at Texas Plant & Soil Lab — other labs utilize similar terms.

Texture — Ranges from 1 (sand) to 3 (loam) to 6 (heavy clay)

Cation Exchange Capacity (CEC) — Texture determines the CEC. Textures of 1 = 3-8 CEC. Textures of 6 = 30-50 CEC.

Organic Matter (O.M.) — Humus increases CEC. About 3.5 CEC increase for each percent increase in humus. O.M. improves tilth (soil physical condition), water and nutrient holding capacity. The more the better. Ideal O.M. levels for corresponding texture levels (1-6 respectively) = 2.8, 3.1, 3.6, 4.1, 4.5, 4.8.

Natural Extracting (CO_2) — Plants produce natural carbonic acid in the root zone which can be used to obtain nutrient values that are more realistic and calibrates to the plant uptake.

NO_3 (N) — This highly soluble nitrate ion moves easily up and down with water and is a constantly changing value. Plant uptake is rapid. Excess can be toxic.

P_2O_5 (P) — Extracted with CO_2, the amount is reported in pounds per acre for the top foot of soil. The amount reported is available to a crop in a normal growing season.
Responses can be expected below 40 pounds per acre and high phosphorus requiring crops may respond to additional phosphate of up to a 200 pounds per acre tested.

Potassium (K) — Extractable using CO_2. This is the amount available to the crop in a growing season. Readings range from of 80 ppm to 120 ppm for crops with high potash needs. Soil availabilities vary with texture, soil moisture conditions, interference from sodium levels and ratios of Na, Ca and Mg.

pH — The acidity measurement is variable. Most crops prefer a pH between 6.5 and 7.3. Neutral is 7.0, above is alkaline and below is acid. The desirable pH level is a nebulous, dynamic determination that is highly variable.

Electrical Conductivity (EC) Salts — This is a measure of total water-soluble salts expressed as mmhos/cm. EC x 640 = total dissolved solids in ppm.

Salt Cations — Water-soluble cations determined by the atomic absorption spectrophotometer. Calcium is important and should exceed 100 ppm.

CO_2 extractable (carbonic acid equivalent) is the same as the plant root process. Sodium is the main extractable harmful element and should be below 180 ppm. The amount of extractable calcium reserve in the soil is also reported and must be known to properly manage excess salts.

Na (CO_2)/Ca (H_2O) and Na (CO_2)/Mg (H_2O) — These ratios help evaluate salt problems and are indicators of the soil's physical condition for water, air and root penetration. The Na:Ca ratio should be less than 6 for good internal drainage. The Na:Mg ratio should be below 20 for regular crops and below 10 for sugar-producing crops such as melons, citrus, sugarcane, etc.

Rating Guide for Soil Test Reports
(Calibrated by Plant Analysis Uptake)

Nitrate
(NO₃)

1-9 very low
10-19 low
20-29 upper low
30-59 medium
60-89 high medium
90-139 high
>140 very high (caution, seedling injury possible)

0-12" sample
Rated as N lb./ac.
ppm = lb./ac. ÷ 4

Phosphate
(P₂O₅)

1-10 very low
11-19 low
20-39 upper low
40-59 low medium
60-79 medium
80-119 low high
120-199 high
>200 extremely high (micronutrients may be tied up)

Rated as P lb./ac.
ppm P x 2.291 = P₂O₅

Potassium (K)	H_2O		CO_2
	CO_2	very low	1-59
	40-59	low	60-79
0-12" sample	60-79	medium	80-99
lb./ac. = ppm x 4	80-99	high medium	100-119
	>100	high	>120

CO_3 (free carbonates)	Mostly Ca and Mg
	0=none
	EH=extremely high

Calcium Ca ppm	H_2O		CO_2
	<19	very low	<149
	20-69	low	150-249
	70-89	marginal	250-399
	90-119	medium	400-599
	>120	high	>600

Magnesium Mg ppm	H_2O		CO_2
	<9	very low	< 39
	10-12	low	40-59
	13-14	marginal	60-79
	15-17	medium	80-99
	>18	high	>100

EC Salts mmhos/cm		
	<0.49	very favorably low
	0.50-0.99	low
	1.00-1.59	slight accumulation
	1.60-1.99	medium little problem — caution
	2.00-3.99	high — affects many crops, treatment needed
	>4.00	very high — affects most crops, treatment essential

Texture	Rating	CEC	Example
	1	3-8	Sand
	2	6-12	
	3	10-20	Loam
	4	15-25	
	5	20-35	
	6	30-50	Heavy Clay

Sodium
Na ppm

H_2O/Na should be over 50% of the CO_2/Na so it can leach through the soil profile. The solubility of the Na is affected by sulfur (acidity), soluble calcium and biology.

When the H_2O/Na is over 50% of the CO_2/Na and the EC (total soluble salt) is high this indicates that better internal drainage is needed and subsoils need testing. When CO_2/Na is high (>180) and the H_2O/Na is less than 50% this indicates a need for chemical and/or biological treatment to increase soluble Na so it will leach. Also, test soil for soluble (H_2O) cations, especially Ca and Na, to determine best salt treatment and management.

Questions for a Grower

Iteration and reiteration line the road to understanding for field consultant and grower alike. At a certain point in the consulting process, Chandler likes to pause and review the lessons with his client. That is the task of this section as well, written as an open letter to growers. The questions are mine, the answers are Chandler's, and they are so valuable they ought to be weighed out on a jeweler's scale.

What is a Soil Test?

A reliable soil test includes three parts:
• Proper sampling, with field and cropping history, and yield goals.
• Chemical tests for nutrients available for crop growth.
• Reliable recommendations for nutrients and amendments needed to attain desired yield goals.

An Interpretation
of a Soil Test for Salts

Field	Text.	O.M.	CO₃	pH	Salts EC	lb. per ac. NO₃	P₂O₅
1 North 1'	3	0.35	0	6.8	0.77	74	99
2 " - 2'	4+		H	7.5	0.67	12	23
3 " - 3'	5-		VH	7.5	0.77	16	9
4 " - 4'	5-		VH	7.6	1.01	15	7
5 South - 1'	5-	0.30	0	6.5	0.53	25	61
6 " - 2'	5+		tr	7.1	3.12	20	4
7 " - 3'	5+		H	7.4	2.06	32	6
8 " - 4'	5		H+	7.5	1.73	31	10

This soil sample was taken after several deep flushing rains and a long history of regular sulfur use.

North Field: There is some sodium salt accumulation in lower levels of the North samples. Note that some P leached to lower levels.

South Field: EC is higher only in the subsoil here due to leaching of soluble salts. Sodium is higher in the subsoil — both soluble and extractable. Soluble H_2O Ca is part of the salt and indicates some internal drainage is occurring at the 4' level with obstruction at the 2' level.

SALT CATIONS - PPM								Ratio	
K		Na		Ca		Mg		Na /	Na /
H_2O	CO_2	H_2O	CO_2	H_2O	CO_2	H_2O	CO_2	Ca	Mg
34	50	90	142	72	396	15	65	2	10
27	71	108	164	73	1744	14	206	2	12
22	59	132	192	61	2240	14	210	3	14
28	61	148	232	72	2136	15	200	3	16
20	48	107	198	86	276	20	46	2	10
13	22	281	352	112	516	26	91	3	14
18	23	267	360	102	1816	27	265	4	13
17	29	252	338	87	2000	25	271	4	14

Extractable (CO_2) Ca is low at the surface. However, above 200 ppm is adequate for regular sulfur use. Deep chiseling can bring more Ca up from the subsoil for a while, and then gypsum or lime is desirable to maintain soluble Ca to aid physical condition for Na leaching. If the subsoil is low in Ca, lime plus sulfur is preferred over gypsum. Other soil treatments of humus type products, soil inoculants, etc. plus energy (sugar/amine) and food (humus) can improve the soil tilth (condition) by increasing soluble Ca from biological activity.

Why Use a Soil Test?

To get out of the guessing game of trying to predict what balance of plant nutrients is needed to attain the most economical yields and the best crop quality. Any test, no matter how rudimentary, is better than a guess.

A fertility soil test can help take the guesswork out of soil fertility and plant nutrition programs. Soil testing can be a precise tool or a ballpark estimate. Any test beats a guess! Why not get the most from time and money spent on a soil test?

A soil test can be for texture — sand, silt and clay. A soil test can also be for biological populations.

Facts:

A. A lab test only reports what is in that sample.

B. Is that sample representative of the entire root zone of the plants in the area?

C. Precise results come from a precise, representative, composite sample from the major root feeding zone, as well as from the deeper soil conditions that affect long-term growth.

D. If you can see a difference in color, texture, slope or plant growth, it will test differently.

E. Choose a lab with experience and knowledge in testing, interpretation, and growing for better results.

F. Only plant nutrient analysis can tell what the plant actually gets from the soil via the roots.

Testing Subsoils

While most plant roots feed at the 4-12" depth, the next 12-24" may also be a major contributor. For the most accurate soil test recommendation, sample the topsoil (0-6" or 0-12") separately and then the next 12" increment of the subsoil. For problems such as germination failure, only the shallow 1-2" root-zone area near seed should be sampled, with an additional slice of the immediately underlying subsoil. For trees and deep-rooted crops, go to 3 or 4 feet, or parent material in 1-foot increments. Subsoil tests evaluate sustainability (mining or building) for long-term performance. Test top- and subsoil samples annually to determine if mining or building-up of subsoils is occurring. Test to 3 feet and 4 feet at three- to six-year intervals to check on the deeper rebuilding progress as well as the soil's physical condition. For better interpretations and recommendations, with each sample please include a brief written history of recent fertilizer use, problems, and yield performance so we can better interpret what the lab test means in your specific situation rather than using general standards. Sample problem areas separately to be compared to adjacent normal areas. Please list future crops and yield goals and a copy of previous tests if possible.

When is the Best Time to Test?

It is never too early or too late to test the soil. Yesterday is almost always the best. A soil test inventories soil nutrient availability for future growth and evaluates nutrient status at the time of testing. The better the accompanying information, the better the recommendations will fit your exact soil situation.

Types of Soil Testing Methods

The CO_2 extraction method works well with all types of soils. The Atomic Absorption Spectrophotometer/ICP and other instruments accurately determine available water and extractable CO_2 minerals in the soil. These natural methods, along with a detailed soil history of treatments, yields, and goals allow for more reliable recommendations.

Chandler offers no retreat into the technology of yesteryear. Quite the contrary, his clients often rely on global positioning devices for proper field mapping. Proper instrumentation enables growers to factor in soil variables. If the devil is in the details, as folk wisdom has it, then so is salvation. The last is to the field agronomist what signs and symptoms are to the physician. In the last case, the goal is human health. In the first case, the goal is profitable crop production with balanced nutrition to feed human beings. Yield goals are not seat-of-the-pants stuff, as far as Chandler is concerned. They are calibrated much as each step along the way is calibrated to physiological needs for healthy plants.

You learn a lot about humates and humic acid use when you get to know Chandler and his associates. Only a few decades ago, T. Senn of Clemson University was almost alone in recommending what conventional agronomists called "foo-foo dust." Innovators ratified Senn as much as Senn ratified the innovators, one of whom was Chandler.

"Crops create new wealth," Chandler says. One consequence of abusing this fount of wealth is salt buildup, a tragedy that Chandler seldom lets farmers forget.

Dealing with Salts

At some point almost any soil report for the Rio Grande Valley addresses saline seeps and those chlorides, nitrates and sodium salts that, in excess, spoil the landscape and annihilate the microorganisms. Killer salts migrate with water, surface with caprice, and haunt the careless composite samples along Highway 77 from the Gulf of Mexico westward to the Pacific Ocean. The best way to manage salts is by determining the ratio of carbon dioxide-extractable sodium to soluble calcium and/or magnesium. This Lengyel/Chandler bulletin is based on field experiences and states with solid finality that a ratio of sodium:calcium approaching 5:1 is barely acceptable.

A greater index indicates a very poor physical condition for water and air penetration to leach the soluble salts. The Na/Mg ratio should be less than 20, or less than 10 for sugar crops, to improve soil structure and tilth conditions required for soluble calcium or magnesium cation exchange capacity. A successful remedy depends on a slight improvement of about 3.5 CEC for each percent of humus. Improving the humus fraction of organic matter is the best way to improve the standard CEC analysis. Base saturation depends on extracting the strongly-held and solubly-held forms of potassium, sodium, calcium and magnesium cations. Lab numbers alone do not provide definitive numbers if the problem is caused by poor drainage. Values achieved seldom calibrate to the plant uptake. For this reason, Chandler's lab determines the level of water-soluble salt cations, most of which are plant nutrients. In any case, water-soluble cations are determined separately. By using the water-soluble along with the extractable values, it can be determined if treatment is needed. Water-soluble calcium (H_2O/Ca) is the key to leaching problems. Extractable sodium (CO_2/Na) is stuck to the soil particle. The ion-soluble calcium exchanges on the soil particles with the extractable sodium, thereby converting Na to a soluble form. Humus and organic matter for tilth are also most important in salt management of cattle and poultry manure. Humates, humic acids, microorganisms and soil inoculants thus arise to assert an intelligence recently surfaced from the underworld of agronomy. Hydroxyls and polymers should not be overlooked by the venturesome grower faced with salt management.

Soluble calcium moves in the bovine with soluble sodium, therefore treatment with chemicals such as sulfur should be used in moderate amounts, counsels Chandler, and this needs to be done several times a year. Deep soil water drainage outlets are essential to the leaching process.

Beyond Plant & Soil Testing

When the tide rolls in, as the saying has it, it lifts the rowboat as well as the ocean liner. Yet the workaday business of sampling the soil, the leaf and the petiole is not half of Chandler's reason for being. He also expends much time and treasure trying to get operating information online. "We need people to help the farmer understand that tools are available and that plant nutrition management answers to the stimuli available. This is not the offer of a magic bullet, nor is it the alchemist's promise to turn base metals into gold." Even without those fine-tuned petiole/leaf tests, Chandler's growers know when they're on track. Insect pressure disappears.

Chandler believes his personal extension effort is on the right track and moving straight ahead. His daily contacts tell him the operators of smaller farms in the Midwest are opting for balanced nutrition, not just for corn and basic storable commodities, but for the specialties that baffle the bureaucrat when he tries to find out what they are.

Hippocrates practiced medicine with 300 natural herbs. The American farmer is now growing twice that many crops, if herbs and medical botanicals are included. All answer to the grand principles that Esper K. Chandler has in his working papers and, more directly, what he ladles out as field advice.

Three decades ago, merely mentioning natural/organic products and biologicals made most chuckle and twitter. No more! On the speaker's platform, Chandler enumerates them all. He details the effect of the foliar route during drought and high humidity. He teaches farmers how to rely on humates delivered through the pencil-thin drip-irrigation lines. He counsels meat protein producers and dairy farmers so that veterinary bills are almost unknown and five- and seven-way vaccines irrelevant. Pesticides hardly rate a mention when nutrients are in place and regularly available, because insects are a disposal crew. They are summoned when needed to remove substandard crops from the scene — Darwinian style — and they are repelled when not needed.

"You don't move back to the fundamentals, you move ahead to them," Chandler reminds. "This chemical agriculture is waking up to the fact that NPK and rescue chemistry are not going to solve their problems." Thanks to the adjuvant inventory of regulators, bio-products, surfactants, the most dyed-in-the-wool "scientific" farmer circa 1950 now finds him or herself tiptoeing back to what was once considered never-never land, to what is now recognized as a way out of the conventional impasse.

Sourcing Fertility

There may be many troublesome words in modern agriculture, but none glows in the dark as much as "conventional." How a recently minted term such as "conventional" came to label practices less than 60 years old in agriculture, a practice that goes back 10,000 years, surely must puzzle etymologists. Nevertheless, we appear to be stuck with the word, and might as well examine it in the context of modernity.

Calcium (Ca)

"On any soil on any continent," says Chandler, "we are required to look at the available calcium level, which is possibly the most variable of the elements required for crop growth." The needs of the soil dictate the kind of lime to be applied. Some soils have fairly good magnesium. Therefore, calcareous sources of lime ask for evaluation. There are many such sources, Chandler cautions. He cites oyster shells, even caliche, and if magnesium is deficient, then high-magnesium limestones are needed.

What, then, is dolomitic lime and what is high-magnesium lime? In Texas, dolomitic lime is notoriously absent, but there are natural lime deposits of 8, 10, and 11% magnesium. "Get to know your highway department because they know where the deposits are," farmers are often told, and it's valid advice. "In Texas," Chandler explains, "most of the local limestone is used in roadbed construction. The Texas-Louisiana Aglime and Fertilizer Association, supported largely by the Sneed family of Georgetown, Texas pioneered the agricultural lime business, supporting research, education, and quality programs for generations. But then when we go to agricultural lime, an overriding factor is the fineness of the grind." Specifically, it must be ground as finely as talcum powder if it is to be reactive.

This is a basic problem. Chandler learned back in his experimental station days that Arkansas dolomite lime ground only to the fineness of beach sand would have little to no effect on pH. This meant more fineness was needed. Such a grind takes a generation of weathering to be useful. But lime with the consistency of talcum powder goes to work quickly as microbes break it down to help neutralize the soil's natural acidity.

The classes and sources of limestone are too numerous to catalog in one sitting. Just the same, mere consideration of the subject makes it necessary to determine what is available to the plant. Or, as Chandler recites, "You have to ask the plant whether indeed it is getting the calcium."

The Rio Grande Valley has soils with 4,000 to 10,000 ppm cation exchange capacity calcium, based on conventional testing. Still, calcium-deficient crops grow on that soil. The lesson is clear. Without the required amount of calcium available, how can such an overload be released? The problem is staggering in its dimensions. Soils well endowed with calcium often produce hungry plants because that prince of nutrients is not available. Conventional agriculture asks for water solubility, and yet the natural product often is not water-soluble. In any case, conventional testing does not look at water solubility (available H_2O/Ca).

That's the why and wherefore of calcium, the major building block of all life. Many soils are not well endowed with calcium. Even if measured, the calcium is often not available. The one lesson commercial farming has to face tells us more than a lot of farmers want to know about those microbes. The microbes alone can slowly regenerate the fertility of the minerals. The business of extracting the minerals is one of nature's finest accomplishments.

Humus and calcium have an intractable partnership in good soil tilth. Calcium is often called the VIP of minerals. Calcium takes top billing in some lab recommendations after humus.

Here, gypsum enters the fray. Chandler puts it this way, "The overriding factor in getting calcium available is converting it to an available form of calcium-sulfate as gypsum. Enter the sulfur content coming from the natural degradation of organic matter, which contains the sulfur nutrient, or it has to come from elemental sulfur itself."

Field/Depth	CO$_3$	pH	Calcium	
			H$_2$O	CO$_2$
North — 1'	0	6.8	72	396
North — 2'	H	7.5	73	1744
North — 3'	VH	7.5	61	2240
North — 4'	VH	7.6	72	2136
South — 1'	o	6.5	86	276
South — 2'	tr	7.1	112	516
South — 3'	H	7.4	102	1816
South — 4'	H+	7.5	87	2000

Chandler prefers to measure and evaluate calcium four ways.

Magnesium (Mg)

The process of loading the soil colloid with an available form of calcium is crucial to cell life. Soil biology is a major factor. From there, the equation progresses to the magnesium factor. Magnesium is not as major an element, but it commands a ratio and has an essential function. Thus, the hunt for sources of magnesium calls up Epsom salts. Magnesium sulfate is a primary source, as is the mined naturally occurring mineral called Sul-Po-Mag and K-Mag, which is sulphate of potash magnesium. When conventional agriculture made its case, the rush was on to buy up either magnesium deposits or vulnerable competitors. Magnesium is a finite resource. This reality had smaller companies finding and exploiting smaller veins, now dominated by the marketing name of K-Mag.

Potassium (K)

The next major fertility requirement is potash, often available as K$_2$O. Most sugar crops require more potash than nitrogen. Potash is a natural, mined mineral. It is not a rare earth, but it calls for entrepreneurial skill and dedication to wrest it from the earth. Deep-vein, hot-water mining in Canada has placed high-cost extraction in the United States onto the back burner. "I'm told that we have some deep deposits of potash in the Rocky Mountains. It can be mined using Canadian technology," Chandler points out, citing the method that has been proved.

Chandler is more than a little concerned about the future of fertilizer inputs, not because of technology that powders, prills, and otherwise refines the materials for field distribution, but because sources are finite and use is often wasteful. "We have natural deposits of potash," Chandler reminds, "from the desert and Dead Sea areas where it has accumulated due to a natural distillation process. These materials have ample amounts of other minerals as well. So, there are many sources of potash, but it is the economic considerations that usually prevail."

As an aside, Chandler explains the range of the subject to natural/organic folks who often want to prohibit the use of potassium chloride because of the chlorine. This, in excess, is a problem. Unfortunately, "you run the cost up to the natural/organic grower when he or she has to turn to more costly sources of potash," reminds Chandler, "and potash is one of the largest quantity elements necessary for production of all crops."

Some soils are well endowed with potash as measured by almost any laboratory inventory, but is it available? And even more important, how can it be made available? To ask these questions is to suggest the availability of an answer. Here is where natural/organic insight comes to the rescue. These denizens of the academic underworld forced down the throats of academia the come-lately consideration of humus, the food for microbial balance in the soil for release of minerals and plant nutrients.

A natural mineral that once figured in the research of William A. Albrecht is langbenite, more commonly known as Sul-Po-Mag, and now as K-Mag. Chandler has extensive experience with this mined product. Langbenite is the basis for both chemical and organic agriculture. It is, as the secondary name implies, a balance of sulfur, potash and magnesium. Unfortunately, many of the owners of such mines have relegated K-Mag to the back burner of company economics. This means little or no investment in production facilities. Cargill controls both ends of major production. As it stands, fertilizer fabricators literally synthesize the natural product much as they do all salt fertilizers. The label misleads farmers who often burn crops because they believe the label, thinking their purchase is the real thing.

Most of the firms that control mineral resources are busily consolidating, as the saying goes, "into a few strong hands." The process erases competition, establishes administered prices, and relies on the old iron law of "What will the traffic bear?" Chandler sharply defines secondary minerals as absolutely essential, the primaries being NPK.

K. ("Potash") Chandler seldom drops the fertilizer subject or the laboratory equivalent thereof without a word on potassium. Potassium is the largest cation in almost any plant. It usually accounts for more pickup than nitrogen. This appears to be a strange statement since nitrogen has the reputation as a dominant element. There seems to be a natural antagonism between potash and phosphorus. The two are constantly trying to tie each other up, and nature loves balance as much as fecundity.

Now the sequence becomes clear. That overload of phosphorus supplied at the beginning of the year tends to run out. Electrical charges figure, most notably the penchant of potassium to tie it up. As phosphorus uptake falters, so does yield. Small amounts of phosphorus in the drip line along with humic acid doubles the phosphate uptake. Moreover, merely using humic acid can deliver as much phosphate to the petiole as a smaller amount of phosphate alone. The two together seem to double the P uptake. This achievement, faced off against the usual research-proven 5-15 percent P uptake, confers an efficiency on precision agriculture only wished for by staid conventional farmers.

Chandler asserts that the above procedure with seaweed hormones and soil inoculants, all together, have quadrupled the effects of available phosphorus. When phosphorus is taken up, so too climbs the uptake reading of nitrogen and all other nutrients. Now Albrecht's sage observation kicks in. Plants in touch with exchangeable nutrients have the capacity for manufacturing their own hormone and enzyme systems, which are needed to challenge insect predators and crop diseases.

Nitrogen (N)

The nitrogen industry demands consideration, especially the synthetic nitrogen industry. During Chandler's early career, he was actively involved in promoting low-cost nitrogen, especially anhydrous ammonia. The name of the game was converting anhydrous ammonia with sulfuric acid into ammonium sulfate, or with nitric acid into ammonium nitrate. Concentration makes urea an end product. Urea has the reputation of being equally acceptable to sustainable or conventional agriculture as 46 percent nitrogen.

Naturally mined sources of nitrogen, chiefly guano deposits of Chilean nitrate, were the main sources of nitrogen before World War II. There was little production of ammonium nitrate before conversion of the munitions industry to anhydrous production. Then came the conversion of explosives to agricultural uses and, in our times, the readily available ordnance for insurgents and terrorists.

During the early years of the American republic, there were more than 1,700 sailing ships importing guano to the East Coast to feed tobacco plants, a major source of new wealth from our nation's greatest natural resource, the soil. Chandler had his exposure to mining bat guano in his early consulting days.

Chandler has his insight to lean on. "What we're doing is concentrating our mineral sources. Ammonia nitrogen exists in the natural cycle of nitrogen. Some 80% of our atmosphere is nitrogen. The gas goes up, condenses, and comes down in rain, and, as fixed by lightning, it becomes ammonia. Conversion takes place with the assistance of various gasses to produce ammonium nitrate and ammonium sulfate in rainfall. The cycle is quite natural." Chandler often pauses as he unfolds knowledge confirmed by his own operation as well as by textbook assurances. He recalls his grandfather referring to summer showers as a dry rain; when accompanied by lightning as a wet rain because the pastures would then turn greener. "There is a limit to what the natural cycle will do. This reality has caused technology to concentrate on achieving quantity at the cost of production with a maximum markup."

Industrial agriculture's objective is to deliver more than nature is prepared to yield. For this reason, the technology has been abused by misuse. That said, Chandler moves to the other side of the question. "In a natural/organic approach, we're trying to balance the equation by using sustainable and environmentally friendly and sound agronomic ways."

Excess seems to be the original sin. Excess consumption disturbs the metabolic balance of the human machine. Excesses in the soil disturb the life cycles of unpaid microbial workers and invite life forms that live on toxicity to have lunch. Now comes the quick fix, chemicals of organic synthesis with consequences that can't be measured entirely because they are not clearly understood in terms of generations still to be born.

Urea

Urea is a natural product with a carbon bond. It can be synthesized, and is. Fully 99% (or more) of urea used is chemically synthesized whether so designated or not. The process starts with anhydrous ammonia, takes carbon out of the atmosphere, then combines it back with anhydrous. In academia, it might be considered either organic or synthetic, but in reality it is produced by a synthetic process.

Phosphorus (P)

Commercial agriculture has yet to face the fact that one of its most sacred major nutrients, phosphorus, is a finite resource. The biggest source over the past 70 years was the sea deposits of Florida. The product was called rock phosphate. It was a high-grade phosphate that yielded excellent results in manufacturing wallboard and superphosphate fertilizers. Armand Hammer with Occidental Petroleum put together his empire on the basis of this rock. Hammer saw the benefits of those high-end reserves, for which reason he assembled them into his portfolio. At the same time, Russia was producing anhydrous ammonia at a low cost. Hammer traded high-grade phosphate for anhydrous ammonia. This upset America's industrial balance, spending an American resource that should have been retained as a treasure. As a consequence, the high-grade rock deposits are largely gone, and the phosphate we are now producing is marginal. The world's greatest supply of high-grade rock is now in Morocco. The 0-46-0 online today is not the same chemically as the 0-46-0 produced years ago. The same is true for 18-46-0 and phosphoric acid. Minute changes in the chemistry tamper with availability to the plant.

Most of the research with phosphate was done with the early 0-46-0. Today, little research is proceeding with phosphoric acid liquids of 11-47-0 and 10-34-0. Phosphoric acid no longer has the chemical characteristics of the product used in early landmark research. What does this do to the availability of phosphorus to the plant? What can the plant tell us?

Some soils are better endowed with phosphorus than others. The $64,000 question that the field observer has to ask is, "but is it available?" How can this nutrient bank account be made available? Perhaps it can't be made available. If phosphorus reacts with iron and aluminum in an acid soil, then this presents a truly monumental problem. Under such circumstances, it may well be tied up until the next Ice Age and beyond. Too much attention is paid to manure-based phosphate as a source of water pollution. We need to learn how to manage it with true natural science. All phosphate reserves are declining. This includes those in Tennessee, Utah, Florida and North Carolina, as well as those on specific islands. They are all declining. Iron and aluminum phosphates are not considered of value to agriculture. On the other hand, excess calcium will tie up phosphate. From Chandler's chair, the heart of the program is to steady the availability of phosphate to the plant in terms of uptake.

"Plants eat every day," Chandler reminds all those who will listen. "When we test the sap of a plant and find that the phosphate drops off, we know that the roots are not functioning properly. Now, how do we stimulate that root to go after the phosphorus that's out there? Now we see the connection to things outside the NPK arena. We use humus and microbes and hormones to get the recovery of that phosphate held in escrow since it was put on years before."

Newly gained field knowledge notwithstanding, the university system still recommends putting phosphate in the soil before the crop is planted. Chandler is considered a heretic by some, a pioneer by most. Nevertheless, he never tires of pointing out that of all the phosphate delivered to the soil as fertilizer, modern agriculture utilizes only 15 percent, often less. It is a steady incursion into the precepts of sustainable agriculture that is breaking with the past. Some of those precepts were uncovered by the world's scientists at the beginning of the last century, and they have been expanded ever since. Chemistry ties up neighboring elements. Soil life breaks the bond.

One fact that sticks like an undigested bone in the throats of conventional farming literature is the clarion call of microbes and humus to the rescue. "We can break the calcium off the phosphate the way we did when we mined the rock phosphate and turned it into high-analysis fertilizer via acidification with sulfuric acid," Chandler explains.

The profit motive has decreed procedures that make nitrogen available outside the biological system. The natural method is friendlier to the plant. The evolution of practices must be considered. Some 40 or 50 years ago, government programs relied on rock phosphate. In alkaline soils, it tied up rapidly. The University of Wisconsin counseled applying sulfur with it. Sulfur, when bacteria oxidizes it to sulfuric acid, releases that rock phosphate to an available form, which is gypsum or calcium sulfate. To illustrate the point, Chandler calls attention to alkaline soils in the Rio Grande Valley. The advice is to apply fertilizer in a band with sulfur or humus products. If the fertilizer is broadcast, the volume is too great for a natural process to release. The chemistry of phosphates has a way of becoming a puzzlement. The early mine-washings of phosphates are now a source of agricultural phosphates. Soft-rock phosphate is a term used for these washings, the theory being that a lot of impurities have been washed away by power hoses. William A. Albrecht was one of the early authorities who counseled the use of these waste materials.

Acidic soils have their way when it comes to the release of unwanted elements. Acidity in the lower pH range not only gives permission for life to burdock, it also enables the uptake of aluminum. Such a result is not available in alkaline soils. When locked up and unavailable, aluminum is not a production problem. When activated, aluminum acts like a claw, grabbing phosphate and locking it up. Much the same is true with iron, this for an eternity.

In spite of the fact that mines are running out of deposits, phosphate has to be considered a renewable resource for the grower. Actual removal in the cropping system is small. Most of the phosphate removed is in the seed. The vegetative volume retains its uptake and holds it ready for recycling. This is not to say that taking of silage is less than mining the soil. Animal agriculture thus comes to the rescue. It recycles the phosphorus through manure, bone and meat protein. Most of the removal is in the bones. Bone phosphate has been canceled out more or less by the Mad Cow disease panic. Honest science should cancel popular fears, but this is not the case as long as folklore trumps science and research.

Mad Cow disease probably comes back to the balance of natural and chemical intervention. The sanctified use of Phosmet along the spines of cattle to battle warble flies in England, grubs in the United States, and buffalo flies in Australia is well known. The late British farmer and lay scientist Mark Purdey made his findings a matter of record, so it appears that transmissibility is more fiction than fact. Chemicals that invade information-bearing genes literally ask researchers for an interview, but the rejection remains absolute because it offends the conventional wisdom called bovine spongiform encephalopathy (BSE). Even the infectious agent has yet to be found.

Chandler puts it this way. "We have these types of diseases because the cells cease to function. Any weakened cell is a repository and an attractant for diseases. Insects and pathogens thrive on toxicity. You can't ever be so right that you can't be wrong." To make his case, he calls attention to Mennonite and Amish farmers who are using the ecologically sound agriculture taught by AgriEnergy in Princeton, Illinois. "I see more and more reports that their animal agriculture requires no vaccines. They don't need pesticides to control fleas, lice or flies. Once they strike a balance with phosphate, it ceases to be much of an input cost."

What Chandler is advocating here is collaborating with the total ecology. The business of balance and health indicts partial and unbalanced fertilization, natural/organic or synthetic. As Chandler puts it, "those who consume healthy meat and milk will be healthier." He shrugs his shoulders, ". . . if we don't destroy the food with processing."

Plant by plant, the pattern repeats itself. A phosphate level that starts high usually runs out at the wrong time. Chandler uses his protocols to keep the level high, thereby pushing 90,000 pounds per acre of watermelons, six bales of cotton per acre, etc. Most consultants, farm suppliers, and growers seem to ignore the obvious, even though academic research reveals that early uptake of phosphate is minimal but highly essential, and at seed, and fruit- and boll-making time the appetite is voracious — even early, apply P only as needed. There are other deficits to early bulk feeding, if a metaphor can be excused. Under acid conditions, iron and aluminum tie up the fertilizer. If pH is high, then calcium ties up phosphorus. The tieup is slow to release. "One Oklahoma research wheat field, fertilized for three years with single superphosphate, gave up residual phosphate for 11 years after fertilization ceased," Chandler recalls.

Each passing year — however green the landscape — few ever realize that it takes a surgeon's precision to sidestep disaster, or at least debilitating returns at every step. Take onions. Chandler gives his protocol to growers with alarming brevity. The prologue is the same for almost all other crops, a soil audit and the remedial measures described elsewhere in this book. It reminds growers to be wary of inserting too much of the basic nutrients into the soil before planting, the resulting deficit being uptake of only 5-15 percent of the phosphorus for utilization. Chandler has demonstrated weekly on thousands of acres for years that spoon-feeding phosphorus according to a crop-logging program of petiole and leaf analysis can result in 50-60 percent recovery of phosphorus with proper adjuvants, and that such a procedure will pay handsome returns on investment for a comprehensive soil audit as well as routine petiole tests.

Beyond NPK

NPK combinations range from those friendly to the livestock in the soil to those that Chandler would not recommend under any circumstances. The most overlooked part of the equation is the humus concentration and the humates. They monitor the effectiveness of the fertilizers called conventional. Add the biologicals to any and all of the above comments. The most progressive fertilizer dealers nowadays custom-mix liquid fertilizers to answer soil readouts, and the real innovators now find that entering the realm of natural/organic inputs puts them several huge strides ahead of their simplistic NPK competitors.

Sources are key. It was the matter of sourcing the final mix that opened the channels for Chandler's venture into an arena all but unknown by hardliners and doctrinaire professors. Sources addressed to specific needs put real balance within reach, a balance that could be maintained by asking the plant. Balances took over the industry, but stopped short of asking the plant, the pasture, or the entire crop.

At the time that natural/organiculture declared itself, zinc, iron, copper, and manganese were routinely added. Farmers used two or three sources of nitrogen and two or three sources of phosphate. They used K-Mag, potassium chloride, and potassium sulfate. The last was chemically formulated, of course, sulfuric acid for nitrate of potash, a product that natural/organic growers reject.

Lime

Observation has revealed that lime is the basis of all soil health in terms of calcium and magnesium. Chandler was shown the sterile acid plots at Auburn University where decades of ammonium sulfate had killed all growth, but very green Bermudagrass was encroaching on the edges. This practical experience was contradicted by peer-reviewed literature that said lime wasn't needed.

Dolomitic lime is very hard. Chandler recalls grinding dolomite to the consistency of beach sand, and getting "very little response." The particles were too large to be active, at least not for the generation making the application. High calcium proved to be a softer lime, but still not as available in some areas as its dolomite counterpart. In Texas, high-calcium lime has to be a veritable talcum powder to be quickly reactive, "a lot like sulfur, it also has to be a powder in order for the microbes to break it down for a quick cycling process."

Molybdenum (Mo)

Ever the teacher, Chandler rolls out an endless inventory of information. Take molybdenum, for instance, a rare element in most soils. Molybdenum supports the enzyme nitrogenase. Without that enzyme, nitrogen fixation is canceled out.

Organic versus Synthetic

There is a fine theoretical line between organic and synthetic. Often it is a fortress wall high enough to obscure the horizon. That's how much it separates thoroughly honest people. But Chandler is ever the pragmatist. "You have to satisfy public demand," he says. Chandler works with clients on both ends of the spectrum. He holds that pesticides and herbicides are not indicated in more natural systems, since the objective there is to make those pesticides and herbicides unnecessary or a last resort.

Successful raised-bed growers have no need for toxic chemistry because on such a scale they can balance everything with relative ease. Even approved organic pesticides hardly lure these innovators.

Yet not every distinction is immediately clear. For example, is copper sulfate with lime a nutrient or a fungicide? "It depends on how you use it," says Chandler, emphasizing the crucial issue of levels and doses. "Selenium is essential up to a few parts per million, then it becomes toxic. Often toxicity is in the dose. It certainly is in the case of sodium chloride. It can be a killer. Yet it preserves meat."

There are strange molecules invented by man that have ramifications beyond the immediate utility. Agronomists have their pecking order: the majors (NPK) and the minors (calcium, potash, magnesium, sulfur and sodium). All are essential, though potash overlaps the NPK designation. The micronutrients zinc, iron, copper, molybdenum, boron, chlorine and nickel are a different kettle of fish altogether. "We know too little about their roles," Chandler admits.

Using Microbes

In Central Texas, Chandler puts microbes into the mix to control cotton root rot. He adds substance by using the rule of thumb: one percent molasses for energy and one percent of a high phosphate natural fertilizer plus adjuvants, all to disrupt the ecology of the destructive agent. Critics often complain that one percent is nothing, yet that trace in contact with a germinating seed upsets all of the theories since von Liebig.

"We tend to think in terms of pounds and tons. We should think in terms of a daily diet, because that plant eats and drinks every day," Chandler advises. "About a single pound of nitrogen is all that an acre of plants can properly consume in a day until it is filling out fruit." There is a bottom line to the economics of sourcing. When Armand Hammer of Occidental Petroleum seized control of salt fertilizers, he assumed the role of evangelist for them. All references to what we now call natural/organic were deleted, and two generations of mainstream farmers learned little about the ecology of crops or even the grammar of the subject. The Hammer bottom line made little reference to the problems of production, ecology or life itself.

Ocean Minerals

Ocean minerals are a source of micronutrients and other unexplored traces. Added to the organic compound, "you naturally chelate them," Chandler advises, "and make them more available." Lignosulphates, an extract from the barks of the forest industry, are acid material, and will chelate almost any element, calcium included. The inescapable conclusion is that everything has to go back to natural processes. The post-harvest analysts who evaluate the food we eat have noted that various compounds — the amino acids, the homocystines, the enzymes — all are improved when produced in an organic matrix. With hormone and enzyme systems in place, calcium, magnesium and potash enable the plant to resist disease and extend shelf life. In a word, natural/organic foods are more nutritious. The laboratory says so. "The difference between a fool and a good farmer is one gentle rain," Chandler concludes by quoting his Grandfather's wisdom from the cotton cropping days.

Tilth & Nutrients

The tilth of the soil has command-performance status, according to Chandler, "especially if you inject anything besides N, P, and K." The humus products have to reenter the soil routinely, both the topsoil and the subsoil. Technology, much of it energy from the new wave of the natural/organic movement, has supplied a whole array of products: humus, lignosulfates, hormones, enzymes, mycorrhizae, microbes, biologicals, etc. As these products make their entry among conventional growers, "the earthworms are coming back." The word seems to be out — minimize, or stay away from, harsh chemicals.

Humic Acid

Can drip-irrigation lessons be applied to dryland farming? Yes, Chandler suggests. Side-dressing lesser amounts of phosphate recommends itself. This applies to other nutrients and to the foliar component as well. Minimum disruption of the root system is advisable. "By asking the plant on dry land, we can deliver foliar feed and stimulate that plant to grow much better, again relying on natural products." Chandler returns to humic acids when detailing foliar treatments. Carbon or humic acid in the foliar application — whether the mix be NPK, or otherwise — will rescue the burn. Even sugar or molasses, both readily available forms of carbon, will reduce the physical burn, even from chemicals or pesticides, while increasing their effectiveness.

What We
Ask of Fertilizers

The Fertilizer Gambit

"Two false premises have swept the republics of learning: partial, imbalanced fertilizers, and toxic-rescue chemistry."

— Sir Albert Howard

Even the ancient Greeks knew that plants relied on nutrients in the soil. Hippocrates, the father of medicine, used some 300 herbs and forbs because observation told him that different plants picked up different nutrients, and he related various plant attributes to various ailments. He probably didn't realize that chemicals of the air contributed to plant growth. It took investigators several centuries to reveal that plants took in carbon dioxide and delivered oxygen to the atmosphere.

Jan Baptista van Helmont, a 17th century Flemish physician, performed his famous tree experiment with a willow tree, a tub of soil, and regular water. The sprig weighed in at five pounds. Soil used in the experiment weighed in at 200 pounds. Five years later, the tree weighed 164 pounds. If the tree picked up its growth from the soil, then the tub's contents should have weighed only 41 pounds. The results proved van Helmont hopelessly wrong. After servicing the tree for five years, the soil in the tub lost only two ounces.

Van Helmont groped for answers and in time the problem was passed along to Nehemiah Grew, a scientist. Grew used the newly developed microscope to study the openings on a plant that appeared to be similar to the pores on a human being, the latter obviously for the elimination of perspiration. It came to Grew that perhaps leaf pores were there to admit air. Van Helmont had overlooked air, yet the willow tree had been in contact with air and all its components: volcanic dust, smoke from buildings, and gasses from fires.

Then came an English physiologist and clergyman named Stephen Hales. He conducted peppermint plant experiments. Hales clamped a glass jar over a growing plant — using suitable controls — and he discovered a wilt in short order. He replaced the plant, but kept the old air in the growth chamber. The new plant lasted about as long as the first one. Something significant and wonderful was happening in the growth chamber. Hales died without finding out what it was.

Joseph Priestly made the connection. Plants indeed took in carbon dioxide and served the planet by delivering oxygen. These early probes notwithstanding, soil chemistry in modern terms awaited the 19th century. Nicholas Theodore de Saussure of Switzerland led the way, riding high on the shoulders of van Helmont, Joseph Priestly, Henry Cavendish and Antoine Lavoisier. All had hoped to construct a Table of Elements by separating air into its component gasses. Later it would be revealed that gasses made up most of the plant.

Then came that discordant note called *Elements of Agricultural Chemistry* by Humphry Davy. Davy had credentials and standing, and what he wrote was generally accepted. He proposed that animal and vegetable matter in the soil was the primary food for plants. He believed that elements in the soil acted merely as stimulants. The real food for plants, he held, was the same as that for animals, namely organic matter.

A few years later, Justus von Liebig gave his lectures before the British Association for the Advancement of Science. Von Liebig argued orally and in print that for both man and animal, support came from the vegetable kingdom. Plants, on the other hand, found nutrient materials in inorganic substances. He proved so persuasive that he conquered the republics of learning. Von Liebig's views were those of a chemist who understood nothing about humus. He was educated in chemistry, and dealt with only a fraction of the variables that Chandler was later to find so challenging. The idea that plants had to be fed only three nutrients, NPK, was too beautiful to let go. In a simplified form, this idea came forward whole-cloth, cast into immutable law, and eventually faced the revolt of the very scientists who had supported it. Looking back, Chandler recalled the instant gratification that agriculturalists sought immediately after World War II. "That and quick results literally ignited the fertilizer industry, and for several decades it admitted no challenge."

The ammonia factories of World War II became idle, closed down by cessation of hostilities. These factories represented instant capitalization and a production capacity for domestic consumption. There had been a shortage of nitrogen, and a general failure of the natural nitrogen cycle. The arrival of a plentiful and cheap source of nitrogen tantalized both the producers and the farmers: the former with profits, the latter with an inexpensive fix.

Chandler offers an explaination but not a defense. "Soils were hungry for nitrogen. All of a sudden we have cheap nitrogen available. Anhydrous ammonia was being produced to make ammonium nitrate for explosives. The explosives are still with us. Take ammonium nitrate and put a little organic matter with it (diesel fuel), stick an igniter in the mix, and you have a bomb, a new wrinkle in the insanity of war." (A ship loaded with ammonium nitrate, bound for Europe under the Marshal Plan to help their agricultural production after World War II, exploded at the dock in Texas City.) Increased nitrogen use caused a rapid expansion of ammonium sulfate production that would engulf Chandler's sojourn with the fertilizer industry: Phillips, Best, Occidental, Kerr McGee, American Plant Food Corp. and others.

War surplus factories and a veritable thirst in the countryside invited a mad rush into nitrogen production. Chandler's role in the Phillips 66 organization involved testing, education and promotion on how to use this nitrogen and how to make use of more concentrated forms of phosphates. The Olin Mathieson Chemical Corporation was the leader for that part of the NPK equation.

"The Houston ship channel," Chandler recalls, "is the world's hub of anhydrous ammonia, ammonium sulfate and phosphate. They concentrated phosphoric acid to make high-analysis phosphate fertilizer. You could ship it so much further than the old 0-20-0 based formulas. The industry was guided by the cost of freight. When Olin Mathieson developed that concentrated ammonium phosphate, they could ship it around the world economically. This meant centralized production rather than a lot of small superphosphate factories across the South."

The high-analysis concentrated phosphate war was waged against the low-analysis single, or ordinary, superphosphate (0-20-0) as defined by Vincent Sauchelli in his *Manual on Phosphates in Agriculture*. Chandler got caught up in this turmoil as he entered into the industry. Old-time, low-analysis pioneers in the local manufacturing plants who used natural tankage, offal, cotton seed, and other plant wastes to supply nitrogen helped imbue Chandler with their knowledge of crop performance based on the benefits of single superphosphate. Unfortunately, ammoniation rules required compounders to comply with the 24-unit rule installed by bureaucrats, who, of course, ignored agronomic evidence. Chandler experienced many internal, behind-the-scenes regulatory battles during the industry's evolution as his knowledge of phosphates grew.

The economics of barging freight up the Mississippi and through the intercoastal waterway seemed to add push to the von Liebig finding that plants only needed NPK, while the charge of false premises was dismissed as of no consequence. Along the Houston ship channel one can still see mountains of calcium sulfate, the residue of precipitated gypsum. The fluorine (and now radon) content in that byproduct has added an environmental concern to its use as sheetrock and soil plaster.

A new trend seemed to take hold as the countryside was emptied of people and farms consolidated. Industrial leavings found their way into the fertilizer bag. Mountains of Florida byproduct from phosphate mills sent fluorine-heavy gypsum — formerly distributed in low-analysis, locally formulated fertilizers — into peanut country as a calcium and sulfur source. All of a sudden, "radon reared its ugly head," Chandler recalls. As a consequence, that practice was stopped. But the business of burying pollutants in fertilizers remained in tune with the dreams of avarice, as Washington state whistleblower Pat Martin has demonstrated.

This postwar bow to the legacy of von Liebig does not suggest that the agronomy of the NPK era simply ignored the microbe in the soil. Chandler's master's thesis in agronomy had to do with legume inoculation by nitrogen microbes and lysimeter studies to trace the leaching of legume and fertilizer nitrogen. Government grants made it possible to compare native legume bacteria to commercial inoculants. Even at the heart of simplistic NPK fertilization, the lesson of natural fertility was not forgotten. He also worked with composting gin trash and sugarcane bagasse and with the use of nitrogen on cattle bloat.

In terms of a basic Fertilizer 101 course, it is quite likely that the student will learn that natural nitrogen from the air comes to barely five pounds per acre per year. That's not enough, not when plants need 10 to 20 times that much nitrogen. This fact led early experimenters at Rothamsted in England to use mineral forms of nitrogen on wheat plants. Then, as now, given soils with humus, results were spectacular.

It was known, of course, that nitrogen from organic matter arrived as ammonia and subsequently changed to nitrate. It was not known until 1878 that changes in soil nitrogen were the handiwork of biological agents, agents that Chandler as a student continued to study. Russia's Sergei N. Winogradsky and Holland's Martinus Beijerinck proved that nitrogen fixation took place in the soil without any legumes whatsoever. The essentials of rapid fixation were found to be the absence of available nitrogen in the presence of carbohydrates and lime.

From that moment on, scientific agriculture turned to mined phosphate deposits and nitrates from Chile. Industry said it was cheaper to supply nutrients than to deal with Nature on her own terms. It was a heritage that ultimately led Chandler to making judgments without reference to approving authorities. New questions seemed to arrive one on the hour. Why was there so little relationship between nutrients measured in the soil and the arrival of nutrients in the plant? How, indeed, did minerals move into the plant? The acids involved were oxalic, tartaric, aspartic, acetic, nitric and citric, and they took their turn performing nature's chore.

Potash mines were discovered, and the potash requirement received front-burner attention. Chandler picked up his sobriquet, the middle initial K. (for *potash*), early in his career. By emphasizing K (potash) as a major limiting growth element, this by using the quick field tissue-test, he was able to illustrate the fallacy of the cation exchange capacity soil test. With NPK in place, and the extraction mode settled in favor of strong acids, the business juggernaught took over. Commercial inoculants made their bow, commanded research, and then bowed out into the wings. If native organisms proved capable of fixing nitrogen, they still failed the machine-perfect laboratory test.

Anhydrous ammonia, the synonym for cheap nitrogen, emerged in the 1930s. W.B. Andrews, a microbiologist at Mississippi State College, was a chief proponent. The Tennessee Valley Authority produced the anhydrous cheaply, assuring everyone that a breakthrough in this field of knowledge was just around the corner. The instant sterilization effect was dismissed because nature's fecundity repopulated the sterilized area rather quickly. Withal, it was the presence of organic matter in the soil that made the salts deliver such showy successes. "We didn't look at what would happen a decade later," summarized Chandler. "We burned all that carbon out of the soil, and that's where we are today with hardpan soils having very little life in them."

One of Chandler's associates is AgriEnergy Resources of Princeton, Illinois. Founded by the late Dave Larson, AgriEnergy started using manure extracts as a mix with the rated fertilizers for spraying corn stover to achieve decomposition. The company is on the grow because it shares Chandler's opinion on rebuilding the soil.

Microbes are rapidly becoming the biggest unidentified trend in agriculture not already styled sustainable. These are soil inoculants, according to Chandler. The trades seem to have enlarged the idea of legume inoculants. The soil inoculants contain naturally occurring beneficial soil microorganisms. There is a large inventory of such bottled products on the market, and they are being used in irrigation water and on the plants and soil via commercial fertilizers, all to improve the plant nutrient uptake and the tilth of the soil. Even without the addition of organic matter to the soil to feed the bugs, an explosion of organisms is to be expected.

"If we put humus, microbes and energy into the mix," Chandler explains, "especially when we can compost products, we can correct the shortfalls inherent in the simplistic fertilization that the sustainable people talk about." Chandler identifies Medina as one of the most successful, this in spite of the down-shouting from Texas State Research and Extension agencies. The derisive term, "Hondo Holy Water", damaged that firm during its early years, but today, worldwide sales are robust. For a product to survive with all that rejection afloat serves as a harbinger of the direction that 21st century agriculture is destined to travel.

A lesson of staggering importance can be drawn from recent experiences in organiculture, according to Chandler. "The soil will regenerate, depending on the mineral balance. This takes us back to the Albrecht school of thought regarding minerals and micronutrients. If these are not in balance, then regeneration from rock minerals is ordered up, and that takes a time frame not workable in our commercial climate."

These few thoughts require Chandler and his clients to recall that our soils were formed over centuries. Any 101 course in soil microbiology reveals that the Russians led the way in that field. Their published works asked science to look at microbial action as if to merge von Liebig's ideas on salt fertilizers with the underworld of natural/organics a century or two before the fact.

Can the current generation wait for its soils to recover? In Chandler's view, a low input attitude suggests to some folklore practitioners that nothing is to be brought in, that a natural carbon cycle and a natural nitrogen cycle will prevail only if patience is installed in the equation.

With organic matter at half a percent or less, too much agriculture has become hydroponics in the field. Painting a nutrient-deficient crop a deep, dark nitrogen-green makes a pretty *Successful Farming* picture, but it also invites insects and rescue chemistry. The verdict on hydroponics is "not sustainable." Simplistic unbalanced fertilizers are also unsustainable. The nuances that support the term "sustainability" are not as doctrinaire as promoters of certified organic laws would have the world believe.

"I was born into a natural/organiculture farming system of necessity and progressed to the cutting edge of the fertilizer revolution," Chandler explains. "Now I've been re-educated and moved on to the cutting edge of the new sustainable agriculture." It was the introduction of humate products atop traditional fertilizers that kicked open the door. Humus, microbes, minerals, micronutrients, and traces pitted pragmatic observation against "settled" science, and some of the settled science became unsettled.

"If I had to name one single happening that made the breakthrough in my mind, it was drip-irrigation." Chandler explained. "Drip made practical, side-by-side comparisons in the field possible. The business of watering two and three times a week made possible a phosphate uptake we never envisioned. This phosphorus increase from 5 and 15% jumped to over 50% rapid recovery and utilization just by putting a small amount of phosphorus in the drip-irrigation system — along with many adjuvants such as humic acid, growth hormones, enzymes, and soil microbes." This is now demonstrated in row crop production and even on dry-land situations. In effect, Chandler and friends have learned how to spoon-feed the crop as it makes its way through the season. "And spoon-feed the organic complex," he adds.

The Chandler approach is so revolutionary, it has prompted the fertilizer industry to ask for his consulting services. The flow of information has made it possible to rebuild the organic matter complex in citrus groves, sugarcane fields, banana plantations, row crops, and in hundreds of vegetable, fruit and nut crops. The USDA's Kika de la Garza Experimental Station at Weslaco, Texas, now raises the organic matter complex of conservation tillage soils a tenth of a percent a year — enough to cancel out global warming if practiced universally. This USDA facility was the only certified organic research station, one which the bureaucrats now refuse to fund.

"There's a lot of pseudo-science based on theology on both sides of the chemical and natural/organic fence," Chandler says, wrapping up the subject with a little wry understatement.

Dissenting Voices

The voices of dissent came on as little more than a whisper in the late 1940s and early 1950s. The first question had to do with the weak carbonic acid that roots exude to etch calcium out of hard marble. It made the nutrient soluble, marble being nothing more than calcium carbonate.

Early on, dissenting voices, voices that Chandler was bound to hear and listen to, charged that fertilization posed questions for which answers were tardy in arriving. This is not to suggest that the role of microorganisms didn't figure in those early investigations. The chemists were concerned with more than nitrogen. They discovered, for instance, that there was little relationship between total content of nutrient elements in a soil system and productivity. This led to determination of how minerals moved into plants.

The earliest investigators figured acids from root hairs had a dissolving action on soil particles. Oxalic, tartaric, aspartic, acetic, nitric, and citric acids took their turn in simulating the solvent effect of sap on the soil. This led to the finding that carbon dioxide alone could be assumed to have solvent action. With that, scientists in general abandoned the weak acid digestion concept — too early, many investigators still feel.

The University of Missouri's William A. Albrecht sometimes told his students of a case in plant physiology where, "in the bottom of a container in which roots grew, there was a polished marble in sand. In that polished marble was etched a root pattern distinct enough to be seen." Here, he said, was a plant weathering the fine polished marble to pick up calcium — because that is what marble is: calcium carbonate. "By means of its own carbonic acid, the root was weathering the rock. A waste product of respiration in the root — the chemical acid which the root gives off — was the weathering agent to make the nutrients soluble."

Counting elemental nutrients became the name of the chemistry game even before the arrival of the 20th century. Some academics sought to reduce soil fertility to an exercise in bookkeeping — as indeed, some do today — to supply NPK. The reasoning behind all of this was best stated by F.I. Anderson in *The Farmer of Tomorrow*, where he rejected the idea that other nutrients would ever run out. Anderson pointed to the Chinese, a people who had farmed for 40 centuries.

This controversy prompted F.H. King of Wisconsin to go to China. He wrote a book, *Farmers of Forty Centuries*. King compared fertilization techniques, and he concluded that Chinese, Japanese and Korean use of excrement, muck, canal mud, refuse, and the like amounted to the same thing as Americans using more sophisticated petroleum-treated byproducts for fertility.

The most astute American farmers realized that each farm and each section of a farm presented differing fertility problems. And with that, soil testing got under way. The grand controversies between chemists and what the Association of Official Agricultural Chemists (AOAC) chose to make official would fill many books.

All of them, however, fell back on an 1852 discovery by the British researcher Thomas Way. Way had noted that when soil absorbed ammonia, a corresponding amount of calcium was released to the drainage water. The exchange mechanism, he found, was seated in the clay, a point von Leibig regarded with abject skepticism. Continued study revealed that such an exchange did exist, and that it involved all cation (positively charged) elements. Moreover, it was developed that the total exchange capacity of soil depended on both the colloidal clay and the organic matter.

Modern soil audits are based on the premises that have survived the scientific deliberations detailed above. Do plants take in nutrients in pure elemental form? Or is the form complexed by protoplasm and made organic? How accurately do soil audits reflect the extremely variable conditions that changing temperatures, growth patterns, and physical conditions — tilth, weather, toxicity — impose? After all, the results of a soil audit detail a condition that no longer exists. Moreover, grinding soil samples imposes a condition that never existed in the first place. These cautions present problems, to be sure. None, however, compare to the damage done by institutional business arrangements in taking agriculture the wrong way.

Wars always author new technology. When World War I ended, fantastic industries geared to producing nitrogen for explosives were left standing without a market. The obvious out was to turn them to the task of fabricating fertility, making farm acres a dumping ground for industrial wastes. Salt fertilizers became current coin. It is a matter of record that from 1918 onward use of salt fertilizers was the course advocated by almost all authorities on the continent. The farmer was prompted to use them as a moral duty. In America, the ground was prepared for their use by a final dose of wartime soil exploitation. World War I saw some 50 million acres go out of production in Europe. To balance this world situation, some 40 million acres of virgin grasslands were turned over by the moldboard plow in the U.S. — much of this land in Oklahoma, Colorado, Kansas and Nebraska. West Texas was never fit for cultivation in the first place, and was not successfully cultivated even after the Dust Bowl days until irrigation arrived.

As economic conditions worsened, the farmer grasped at all and any means for increasing his volume so that he could produce more for less. Acid-treated fertilizers and the various nitrogen fertilizers picked up steam going into and coming out of depressions. Potash and phosphate artificials were bagged and sold to kick up the mineral reserves of the soil. In the lawmaking bodies of England and the United States, NPK became holy writ and the profitable background for the NPK mentality.

As expected, this legislation of biology took on a business format in the United States. Institute and association people figured in, but this was only because legislators handed off the ball. Anxious not to make mistakes, the lawmakers called on the Association of Official Agricultural Chemists and fertilizer trade associations who were the forerunners of the National Plant Food Institute, now called the Fertilizer Institute. These experts spoke with a fitting aplomb. The science they presented was thus enshrined into laws because it was current, not because it was final. It never is. Those who objected didn't matter. They lived in the underworld of farm technology in any case, and would not emerge for decades. Chandler took note because the observations could not be denied and because he was involved in the legislation and education of that era.

Model laws and the acceptance thereof may have been generated out of a bad situation. Admittedly, there were rascals afloat in those days — there always are — and the law presumed to put fertilizer quacks out of business. The laws also put out of business fairly respectable competitors who were not sophisticated enough in political matters to live with state government red tape, even though their science was superior. State chemists got into the act because they represented government clout for favored business enterprises. After all, it was to big business that state employees looked for post-retirement employment, and even present professional advancement. Therefore, to preserve their own nest, the fossil-fuel people dominated the plant-food people, and the state employees got together in their white robes to protect the farmers and themselves with a model code. This code called for fertilizers to contain certain high percentages of NPK, and for combination fertilizers to contain aggregate totals of not less than 20% (later raised to 24% at the behest of the industry) of NPK — acidic and factory-soluble.

In the farm press and on fertilizer bags, all of this seemed to say: Protect the farmers. The real message was something else. Ad copy said low-grade vs. high-grade fertilizers, and the implication was always that 20-20-20 was much better than 2-2-2. Presumably a super- or triple-phosphate was better than a low formula affair based on ordinary or single superphosphate. By verbal sleight of hand, water-soluble and available in the *laboratory* was made to mean water-soluble and available in the *field*.

Chandler was exposed to the history of the fertilizer industry while working with Phillips 66 Petroleum, a basic producer of N and P concentrated fertilizer, that sold to local manufacturers, who then supplied farm dealers, who in turn were the conduit for the trickle-down to the farmer. During that era a regimented supply-and-distribution system existed. Chandler's job as an agronomist/salesman was to help the manufacturers educate the dealer and farmer in the wise and efficient use of fertilizers based on soil tests.

That was a time of transition for the fertilizer industry. High-analysis pushed low-analysis into oblivion. A separate revolution of blends versus formulated, homogenous granules was also in motion. Old-time small-plant operatives who once used animal waste, tankage, offal, and blood meal, as well as plant material from cotton-seed oil mills, supplied Chandler with the history of a rock phosphate based industry. It had the early grade of 10-2-2 PNK that was arranged in order of plant needs until the control chemists mandated total available NPK and changed it to 2-10-2. The high-analysis phosphate producers bullied compliance with the 24-unit rule, which forced over-ammonization of superphosphate, thus rendering it unavailable. These old-timers recited their experiences showing the benefits of rock phosphate as a fertilizer. Early studies of the lungs of workers in the rock phosphate mining industry revealed that long-time exposure to rock dust produced healthier lungs. These accounts were etched into the inquisitive mind of the young agronomist.

Ehrenfried E. Pfeiffer was quick to point out that the Mitscherlich vessel wasn't a farm acre, and his criticism of the simplistic NPK idea assured him a place in the underworld of farm technology, from which he is emerging only now.

If a phosphate is declared available, Pfeiffer said in abstract form, it means available with reference to the laboratory test. In farm soil the conditions might be entirely different, and available phosphates might become entirely unavailable — so unavailable, or tied down in fact, that they remain locked up, with only a small percentage going into the plant. In highly alkaline or acidic soils, available phosphates can become entirely unavailable. Indeed, USDA Beltsville, Maryland tracer studies revealed that of the applied phosphates, only two to 10% showed up in the plant, whereas the rest were tied down in the soil: unavailable.

Not that this available-unavailable business meant much to most fertilizer purveyors. Many do not seem to understand that the acid-treated stuff comes from rock phosphate in the first place. In the super-phosphate form, rock phosphate is treated with sulfuric acid. This makes the tricalcium phosphate form water-soluble, resulting in a monocalcium phosphate form. This does more than permit the petro-chemical industry to work off acid byproducts. The chemical reaction involved causes 20% superphosphate to be represented by about 20% monocalcium phosphate, which is presumably desired, and about 55% calcium sulfate, or gypsum, which is the same stuff that some purveyors trade name and sell as a one-shot cure-all. In the bag and in the labora-tory, 0-20-0 is better than rock phosphate, but soil systems are never in a bag or in a laboratory and they know little about solution botany.

There are variations, of course. The designation 0-45-0 is commonly called triple superphosphate. Here the acid is phosphoric. This elimi-nates calcium phosphate in the end product, as does manufacture of the end product called ammonium phosphate. Rudolf Ozolins, the Latvian-trained agronomist, once put it this way: "When rock phosphate is converted from tricalcium phosphate to monocalcium phosphate, this highly unstable form is subject to natural reversion back to the stable tricalcium phosphate form." The rate of reversion differs. The pH, the free calcium in the soil, the organic matter, the iron and aluminum in highly acidic soils — all these things figure in. "I would estimate that 75% of the monocalcium phosphate (di-ammonium) used in farming reverts back to the stable tri-calcium phosphate form within three months. In some soils the reversion takes place within hours," was Ozolins' summary. As soil conditions worsen, release of nutrients from rock phosphate also worsens, and the NPK farmer becomes married to buying more acidic fertilizer, each go-round worsening still further the structure of his soil.

Much the same is true of potash, a nutrient that gave Chandler his identity as the expert on potassium. According to the standard fertilizer guides, potash should be water-soluble. Water-soluble means that the salt goes easily where water moves — into the plant, or away from it. Pfeiffer warned about luxury consumption: forage crops, grasses, and legumes absorbing five to six times more potash than is normal and good for either plants or animals. It would be better, said Pfeiffer, to have a stable form of exchangeable potash from which plants could draw at their discretion.

Plant roots excrete organic acids and enzymes. They in turn favor microorganisms which release more organic acids and enzymes. In a biologically active soil, there is a continuous process of give and take, release and absorption, availability and taking into storage, according to Pfeiffer. As soon as we consider this process, the problem is no longer an excess of water-soluble potash, the importance of having a source of total potash which is made available by way of the biological process.

F. Lyle Wynd of Michigan once put it this way, "The nutritional aspects of soil fertility depend on the activities of living microorganisms and on the electrical properties of its non-living colloidal components. The supply of nitrogen, phosphorus, sulfur, and of many plant foods as well, is completely dependent on the metabolic cycles undergone by these microorganisms."

The question, then, has never been whether to feed the plant directly. The best way is to feed the soil first, which in turn feeds the plants. For, as Wynd says, plants obtain their food from this colloidal and dynamic, this biologically complex organization "in a manner of speaking, only by permission from the soil." That is why the sustainable farmer deals with sandy soils differently than he deals with clay or peat soils. In the first case, low organic matter permits leaching. In peat soils, a low biological process prevents root uptake. In all cases, the answer is to build humus, to mobilize soil life. Fertility that can be added here and now must be geared to that particular soil's exchange capacity. Anything destructive to that exchange capacity is destructive to the soil's life in both the short and the long run. Nothing illustrates the point in question better than a consideration of how nitrogen is to be supplied.

It has been reliably estimated that about five percent of the total organic matter in a soil is present as nitrogen in various compounds. In terms of two percent organic-matter soil, this means about 40,000 pounds per acre, five percent of which is nitrogen. A two percent organic-matter soil therefore has a reserve of 2,000 pounds of nitrogen.

Nitrifying or ammonifying microorganisms transform a small fraction of this storage material to nitrate or ammonia, enough in a living soil to sustain plant growth. If we feed these organisms and otherwise provide for a proper balance, the biological process in soil makes these resources available.

Put another way, there can be from 1,000 to 50,000 pounds of aerobic bacteria per acre in a given soil system. At the upper end of this range, microbes fix enough nitrogen in their bodies to preclude the necessity of ever buying that anion nutrient requirement. It has been demonstrated that non-symbiotic nitrogen fixation can add a Pfeiffer-calculated amount of nitrogen per acre per year, which is more than a farmer can afford to buy. One group of these fixers is the aerobe called azotobacter, operating in soils with a pH range of 6.0 or more. They fix atmospheric nitrogen in their cell bodies. The second type is the anaerobe called *Clostridium*, operating after aerobes have processed the oxygen. Other nitrogen has been returned to the soil via manure and other symbiotic fixers such as the legumes. Here organisms live in symbiotic relationship with plants and fix nitrogen from the air. The wonderful process by which ammonia-nitrogen is converted to nitrate-nitrogen for plant use is a miracle. Yet humans feel that they understand it better than they understand sugar production and energy transformation from the sun.

This has led chemists to develop commercial nitrogen fertilizers and to make salt traffic into soil systems a profitable chemical company enterprise. Ammonium nitrate, potassium nitrate, calcium nitrate, sodium nitrate, ammonium sulfate, ammonium phosphate, and anhydrous ammonia all presume to do quickly what Nature does in her own good time.

But if easily available ammonia or nitrate is offered, nitrogen-fixing bacteria are spoiled. They become consumers and feeders rather than fixers. The farmer on a commercial nitrogen fix thus loses twice. His failure to manage his soil system gives him a nitrogen problem, and his attempt to buy nitrogen gives him additional fertility problems. There is a reason for this.

Much of the potassium available to farmers in commercial fertilizers is muriate of potash, which is potassium chloride. This is a salt, meaning that one atom of positively charged potassium is combined with one atom of negatively charged chlorine, an acid element. In moist soil, salts dissolve. The positively charged potassium atom is separated from the chlorine and attaches itself to the negatively charged clay and/or humus colloids, thus leaving the chlorine free to leach from the soil.

Almost all cation-plant nutrients in so-called commercial fertilizers operate in this way. They are sold as carbonate, chloride or sulfate salts, and they react much as muriate of potash. The meaning here should be clear. It takes humus to make the salts work, but harsh treatment of humus wastes it away. Anhydrous burns away the humus supply just as surely as fire or sulfuric acid applied to organic material does away with the parent material for future humus supply.

The salts have other shortfalls, as tests at Rothamsted Agricultural Station in England have amply revealed. Just one of the thousands of those tests should illustrate the point, ammonium sulfate being the salt in question. Applied to the farm in excessive amounts, this fossil-fuel fertilizer accounted for tremendous reduction in soil bacterial count, and that same salt simply destroys the earthworm population. Indeed, the USDA once published a bulletin recommending ammonium sulfate as a means of killing earthworms, such as on golf course greens. Long-term research plots at Auburn University in Alabama remained sterile for decades because of excessive use of ammonium sulfate. Of all the attempts to stifle natural cycles instead of utilizing them, removal of the earthworm link is the most self-defeating. The Valley of the Nile in Sudan remained fertile for centuries because the environment suited earthworm propagation. Castings from these little critters can add from 10 to 200 tons of prime fertility per acre per year. Earthworms apparently concentrate plant nutrients and make them root-available. The New Zealander, B. Worley, tested the growth rate of plants in boxes with heavy earthworm populations. He found crop production increased as follows: rye, 63.9%; potatoes, 135.9%; vetch, 140%; field peas, 300%; rape, 733%. Other investigators watched grass production improve 271% in earthworm-loaded soil, and increase an incredible 463% when grown in pure earthworm castings.

The chemical logic behind this process has explosive implications. Castings invariably have three times more magnesium, five times more nitrogen, seven times more phosphorus, and eleven times more potassium than similar soils. Earthworms do not like readily soluble minerals. They can't survive in soils without food, namely organic matter. When fertilizers treated with sulfuric acid rain down on this little plowman's home, his life cycle goes into a tailspin.

There are certain dangers associated with using almost all of the fossil-fuel fertilizers. Concentrated superphosphate is caustic and eats the keratin tissue off cattle hoofs when used as a barn deodorant. Potassic fertilizers are also caustic and destructive of biotic life. And they etch their handiwork into the iron of machinery as well. The king-pin in reducing the soil-colloid holding-power is hard nitrogen, a crutch that worsens the limp and ultimately cripples the land. No objections would be raised against salt fertilizers were it possible to use them without long-range deficits or lowered plant vitality. But there are deficits. The entries can be read everywhere. Researchers tell us that super-phosphates, complex zinc, copper and iron prevent their uptake by plants. NPK fertilizers can't be combined with trace elements, not to good effect, in any case. Frequently, they annihilate biotic life. Pouring them into soil systems and then building dams downstream as soils go dead is much like whipping an animal for not having human intelligence.

Yet it is a fact that as soon as organic matter content exceeds two percent in soil systems managed by good sustainable farmers, these soils build up a reserve and draw from it. There is no washing out of potash or loss of nitrogen, and all three major nutrients — NPK — are always available and root-acceptable under eco-conditions.

Land Application of Organic Fertilizers or Amendments

Applying organic materials to your land can add beneficial nitrogen (N), phosphorus (P), potassium (K), micronutrients and organic matter to your soil. Organic materials can increase the soil's water-holding capacity, improve aeration, decrease erosion, and promote biological activity in the soil. Organic fertilizers can be very beneficial to pastures, crops and lawns, but they can contaminate surface and groundwater supplies if applied excessively or improperly.

Preventing Environmental Problems

When too much organic fertilizer is applied to land, plants cannot use all of the applied nutrients. Then, when rainwater runs off the land, it carries these excess nutrients into lakes and streams. N and P occur naturally in streams and lakes, but excessive concentrations can accelerate eutrophication, a process in which dissolved nutrients stimulate the growth of aquatic plants and algae and reduce the level of dissolved oxygen in the water, which can harm aquatic life.

Organic fertilizers may also contain pathogens (disease-causing bacteria or viruses) that can be carried into surface water by rainfall runoff. Pathogens degrade water quality, making it unsuitable for recreational uses and greatly increasing the cost of treating it for use as drinking water.

Selecting the Proper Application Rate

To prevent environmental problems, the most important management practice is to develop a nutrient management plan that bases application rates on annual soil tests and realistic crop yield goals. Fertilizer applications that exceed soil test recommendations waste time and money.

Annual soil tests will tell you the nutrient content of your soil. Then you will be able to determine how much fertilizer your crop needs by subtracting the available soil nutrients from the crop's total nutrient requirement. Remember to base your total nutrient requirement on a realistic yield goal.

Once you know how much supplemental N, P and K to apply, test the organic material you will use to determine its nutrient concentrations.

With this information you will be able to calculate the appropriate application rate.

Texas Cooperative Extension's Soil, Water and Forage Testing Laboratory will analyze soil and organic matter samples and determine the proper application rate for your crop. Instructions for using this service can be found at *http://soiltesting.tamu.edu.* There is a minimal fee.

Remember the following when determining application rates for organic fertilizers:

Apply at rates to meet crop P requirements.

Apply supplemental N to meet crop N requirements at realistic yield goals.

Split organic fertilizer and supplemental N applications so that smaller amounts of nutrients are applied at any one time.

Checking Your Application Rate

Calibrate your application equipment periodically to ensure that it is applying the intended rate. Here is a simple method of verifying your application rate:

1. In an area where material can be easily applied, spread a tarp or plastic bag on the ground and distribute the organic fertilizer over the tarp or bag at your normal application rate.
2. Measure and record the length (L) and width (W) of the tarp or bag used.
3. Carefully collect all the material that was applied directly on top of your tarp or bag.
4. Weigh (M) the collected organic material.

Do this procedure at least three times.

With the area and weight collected from each sample, you can calculate your actual application rate (AAR) and determine whether it corresponds to your desired application rate. Remember that if application rates are to be made on a dry-weight basis, the tons per acre (wet basis) should be divided by the percent moisture content (MC; decimal fraction) to get the correct application rate. The example below demonstrates how to determine your actual application rate.

Example: Calculating Your Application Rate

A grower needs to apply 2,000 pounds (dry basis) of composted manure per acre to achieve the proper nutrient balance for his desired crop and yield. To check his actual application rate, he has set up the test as described above. The tarp he used measured 36 inches wide (W) by 48 inches long (L). The moisture content (MC) of the organic fertilizer is 5% (0.05).

After each application, he weighed (M) the sample and recorded the values. The results are as follows:

Sample 1 (M1): 0.5 pounds
Sample 2 (M2): 0.6 pounds
Sample 3 (M3): 0.5 pounds

Using the following four equations, he determined the application rate for each sample.

<u>Sample 1</u>

M1 (lb.) × 6272640 = AAR1 (lb./acre.) . . . Equation 1

L (in.) × W (in.)

0.5 (lb.) × 6272640 = 1815 (lb./acre)

48 (in.) × 36 (in.)

<u>Sample 2</u>

M2 (lb.) × 6272640 = AAR2 (lb./acre) . . . Equation 2

L (in.) × W (in.)

0.6 (lb.) × 6272640 = 2178 (lb./acre)

48 (in.) × 36 (in.)

<u>Sample 3</u>

M3 (lb.) × 6272640 = AAR3 (lb./acre) . . . Equation 3

L (in.) × W (in.)

0.5 (lb.) × 6272640 = 1815 (lb./acre)

48 (in.) × 36 (in.)

After determining the actual application rate (AAR) for each sample, he then takes the average of all the samples (Equation 4).

<u>Average Actual Application Rate</u>

AAR1 + AAR2 + AAR3 = AAR (lb/acre) . . . Equation 4

3 × (1 — MC)

1815 + 2178 + 1815 = 2038 (lb./acre)

3 × (1 — 0.05)

So, his actual application rate (AAR) is 2,038 pounds per acre. This value is very close to the desired application rate and verifies that his equipment is properly calibrated.

Other Considerations

Location — In order to protect drinking water, fertilizer should not be applied within 150 feet of any private water well or within 500 feet of any public water well. A 100-foot buffer should be left between the application field and nearby streams, lakes or stock tanks.

Weather — Take note of current and predicted weather conditions when scheduling the application of organic materials. If heavy rain is expected within 48 hours, delay the application. Fertilizer applied to wet, frozen or sloping soils is even more likely to be carried away from the application site by runoff or erosion.

Soil — Although nearly any soil can benefit from applied organic material, soils that benefit most are those with good soil depth, no restricting layer in the root zone, a moderate rate of permeability and drainage, and a good nutrient-holding capacity.

Tillage — On cultivated sites, soil should be tilled within 24 hours of application. Tillage breaks up organic material and incorporates it into the soil, which decreases the risk of nutrient loss and places needed nutrients in the crop root zone. Incorporating freshly applied organic fertilizer into the soil also decreases odors.

Conclusion

Several studies have shown that using animal and plant by-products as organic fertilizers is a way to benefit from these resources, rather than wasting them with other disposal techniques. For the most environmentally and economically sustainable program:

Establish a reasonable yield goal.

Apply organic fertilizers at appropriate rates based on fertilizer composition and available soil nutrients (as determined by organic and soil sample testing).

Apply nutrients at the time they are most needed by crops (approximately 20 days after emergence).

Avoid making a single high-rate application (greater than the crop N requirement) because it can degrade water quality, even on sites with low soil-nutrient levels.

Do not apply fertilizer or litter just before a heavy rain is forecast.

To further reduce the possibility of water contamination, alternate organic fertilizers with commercial (inorganic) N.

by Justin Mechell, Daren Harmel and Bruce Lesikar
AgriLife Extension Texas A&M System: http://agrilifeextension.tamu.edu

Bad Science
and Big Business

Men like Chandler make pragmatic observations and record results on forms that professors might correlate into a database if they could, but usually don't because funds are lacking. In 1962 Rachel Carson published *Silent Spring*. She called into question the matter of chlorinated hydrocarbons and organophosphates in production agriculture and therefore in the environment. Her critique has not spent itself, which is why natural/organiculture is growing exponentially year after year. This revolt of the people is mild compared to the popular rejection genetic engineering met with in Europe.

Bad science decreed that weed and insect proliferation was a sort of demonology that responded only to chemicals of organic synthesis. Good science reminded the farmer once on the hour that plants in touch with exchangeable forms of calcium, magnesium, sodium, potassium, and phosphate construct their own internal hormone-enzyme systems.

Good science meant dealing with nature on Nature's terms. As quick-fix genetic engineering took hold, always with an attempt to realign nature, the shortfalls and tragic nature of cross-species propagation at the molecular level became evident, and public debates, moratoriums, and outright bans — as in Greenland — caused the controversy to assert itself in the thinking of clients and consultants alike. As big business money penetrated political circles, not a few scientists with credentials and standing warned that this new tragedy meant the end of "humanity as we know it and the world at large," that this new art and pseudoscience was inherently hazardous.

The opening gun was fired when BST (Bovine somatotropin) was approved for injection into lactating animals in order to double milk production and burn out milk cows in one and a half lactations, on average. Next came the GMO tomato, squash and vegetables too numerous to mention. The big leap in agriculture was soybeans and corn, with other basic storable commodities standing in line for the transgenic treatment. As a rule, ag consultants continued to deal with soil fertility, and the more sophisticated asked the plant. Since labeling of GMOs was forbidden, no one saw the creeping debilitation afloat. A few farmers with microscopes detected a red fusarium mold on corn, and others saw debilitation overtake feedlot cattle and dairy cows. The business of transplanting human genes into animals and animal genes into vegetables offended the ethical sensitivity of most people. Withal, it was the danger of genetic pollution that most offended ethical agronomists.

Those Variable Soils

Testing Variable Soils

"If you discuss soil," says Chandler, "you have to put 'variable' on the other side of the equals sign." During his days at research stations, small 7 x 14-foot plots had a go at forage soil and plant testing. The size of those plots was governed by the availability of space as well as by the requirement for alleyways. The goal was to maintain as many as 120 of these randomized, replicated, and repeated multiple-year studies. The usual procedure was to test soil 6" deep in each replication and plant test each cutting. There might be 21 different treatments in a test with four or five replications. Looking over the plots you could see variations in the way the plants would grow, even though there wasn't a difference in treatment. Chandler recalled the scene this way. "The superintendent, Dawson Johns, insisted that we sample each individual plot separately. We'd take, say, seven cores throughout that 7 x 14-foot plot. We'd test these representative composites from each plot. So here you had quite a bit of variation in a very small area. On that Louisiana State University North Hill Farm Experiment Station, calculations would be based on replicated variations between plots. Dr. Darrell Russell, soil and plant chemist analyzed each sample. Years later, that valuable data was still wallowing in the bowels of University bureaucracy."

Healing Wounded Soil

When Chandler returned from combat in Korea in the early 1950s, he encountered more than a cotton allotment. President Dwight D.

Eisenhower's administration set up a Soil Bank, which seemed to comply with the Committee for Economic Development's mandate to consolidate farms into big units and close down the type of agriculture that existed during the Depression '30s and wartime parity '40s. The erosion left over from ravaged soil invited correction. This meant factual research, aimed at replacing the cotton farmers, turning Kansas wheat plots into mega-fields, and complying with Ezra Taft Benson's injunction to "quit mollycoddling the farmers." Conservation programs planted millions of acres of trees across the South, often on eroded hill land, to build back the soil. Many of those forests are still productive today.

The range of work on that postwar station involved crops, pasture, pine tree fertility, dairy, peach production, beef cattle work, poultry broilers and layers, clearing land and forage production of coastal Bermudagrass hay, which was shipped to the main campus. The farm's diversity goals included silage production, row crops — cotton, corn and milo, as well as grass and legume test crops that seemed to pose questions. There were no bureaucratic limits on what could be researched, and there were no caveats tied to industrial grant money. Chandler was allowed to put in all the test plots he desired, including his own food patch.

Coastal Bermudagrass, developed by Dr. Glenn Burton of Georgia, rated front burner attention because of its potential for closing those soil wounds that wind and water erosion had accounted for. Coastal Bermuda is a prolific grass, one capable of taking hold with deep anchoring roots in places like eroding gullies. The state of the art decrees one treatment regardless of variations spreading across either plots or row acres.

"We would take manure from the dairy and poultry operations and straddle those gullies, usually dumping the manure rather than slinging it to the bottom. With that fertility, the Bermudagrass would grab hold and stop the erosion," Chandler recalls.

"Variable soil profiles are the norm," says Chandler, "even under the best of circumstances." Old red subsoil often canceled out the small amount of organic matter discerned now and then. Very acid soils have little or no nutrition in escrow.

Circa 1950, "we introduced grain sorghum to that parish," remembers Chandler. Every innovation seems to invite an unsought counter-development. In the case of milo, it was birds. The feast of small grains for creatures that like grain made the station look like something out of an Alfred Hitchcock movie. Propane guns with timers failed to stop the birds. "We used firecrackers spaced on lengths of cotton ropes tied to trees so that fire smoldering up the rope would explode the firecrackers at intervals. Next we had to get retired folks to fend off the winged predators with shotgun pellets interspersed with the firecrackers or the birds, mainly crows, would take everything."

The soil proposed and the crows disposed because the environment around Homer, Louisiana made it tough on wildlife. There was a time when it was impossible to lose sight of a nearby cotton field in that environment. By the end of the 1940s, King Cotton was not even a poor pretender to royalty — research on cotton, fumigation, and fertility studies notwithstanding. Small farms became social, political and research anathema. "Don't do research on small plots," the sotto voce admonition. "Use commercial farms. That's where the need is." Chandler and his associates found three commercial farmers still in the cotton business. Today that parish has not one. The soil-mining era having spent itself, the soil would no longer permit it.

Chandler calls his and the grower's nemesis "bad cultural practices." These include excess tillage, wasteful fertilizer and water use, bulk soil treatment when spoon-feeding is indicated, and virtually complete ignorance of foliar nutrients and natural adjuvant application.

Looking at Soil Sampling Differently

The paradigm changed the day Esper K. Chandler moved into the Rio Grande Valley because the soils had changed. Gone were the hills and sand, now replaced by a flood plain in an arid subtropical climate. The Rio Grande originates in the mountains of Colorado, meanders through New Mexico, passes El Paso, and then turns toward the Gulf of Mexico. It is one of the historical rivers of the world in terms of the land nearby and the irrigation systems it supplies. Here, as in present-day Egypt, the plains are spared the inconvenience of floods with a dam, canceling out the nutrient fix and salt leaching that every flood accomplished. With irrigation the salts build up. In Egypt cotton, once a famous staple crop, is almost nonexistent because the builders of the Aswan Dam's irrigation system failed to plan for internal drainage for salt to facilitate leaching.

A soil chemist named Schultz started the laboratory in 1938 that Chandler later made his own. It was the first soil lab in the state of Texas. Schultz developed the four-foot-in-one-foot increment profile. This technique had its faults, the main one being the prevailing concept of the hour. The soil was rich in minerals and equipped with the full pantheon of micronutrients. Everyone believed that all you needed was nitrogen. If you leveled the land, controlled salts, irrigated and used nitrogen, you painted the landscape green. That, points out Chandler, is what we're still doing seven decades later. "We're mining our soils, particularly of organic matter and minerals."

The standard procedure is to take several randomized core samples, then mix the samples to achieve a representative composite. Yet even plot experiments reveal a significant difference within a few feet. Uncommon good sense analysis told Chandler that precision farming was indicated. In time, global positioning systems enabled a precision never envisioned when Chandler was simply observing as a researcher, not a pioneer and visionary.

In Chandler's view, the methods of the natural/organic folks demanded that conventional agriculture pay attention to claims about organic matter, humus, soil microorganisms, and the conversion of inorganic minerals to soluble organic for root uptake. But it was the recognition of variations between side-by-side trees or row crops that exhibited a difference and invited the farmer to address those differences with the use of plant nutrients. Citrus trees were usually 10 to 24 feet apart. A soil sample on one side of the row often varied greatly from a similar sample on the other side. This prompted the marking of trees so that subsequent samples could validate findings and measure the character of every response where sampling variations did not influence the evaluations. The bottom line information revealed that there were more inherent differences in the soil than in the treatments. The uncomfortable conclusion was that most of the earlier basic research was badly flawed because it was not calibrated to plant uptake.

Some of Chandler's mentors wanted to remove many of those inherent variations. Immediately, certain appropriate conclusions started closing the gap between organic folklore and so-called settled science. It was a small step to repair the soil with humic acids or soil inoculants, eschewing the NPK code.

"We came to an inescapable conclusion," Chandler conceded. "We were introducing more variations via our testing procedures than were imposed by the differences we were trying to measure." Statistics don't lie, but they don't digest facts very well either. Peer review somehow failed to square with reality. It was an awesome discovery, this business of methods and materials introducing more variables than were the goal of increased production. Then, as now, "too much of our work was and is theoretical and formula founded, and too much of the practical farm-applied research is funded by people and firms with a product to sell, products that they can protect with a patent or copyright."

The situation has taken more alarming turns than a Roman taxi. Witness Monsanto and its relentless effort to develop and sell Roundup Ready soybeans, glyphosate, GMO canola, and all the rest. It makes dealing with nature seem less than scientific by comparison. To regenerate the soils that have been mined out, "we have first to understand that it is recoverable. It forgives many of our transgressions, but to recapture the values both research and practical agriculture have to obey nature, not the laboratory approximation thereof," says Chandler.

The Vanishing Organic Matter

When the Rio Grande Valley was first cleared for agriculture in the 1920s and 1930s, the organic matter complex ran between 3 and 5 percent. Today, it is hard to find a one percent organic matter soil. In lighter soils, it is hard to find a half of one percent. When Chandler took over the lab in the 1980s, the trend was to quit testing organic matter. At that time, Rodale did a monitored study of testing procedures, and taking retention readings of organic matter readings was part of the science. At the same time, *Acres U.S.A.* published a smaller study that serendipitously asked some of the same questions. It proposed checking on how recommendations differed when the same sample was sent to different laboratories. The results revealed a serious disconnect, no two labs drawing even related conclusions from the same samples.

It was the prospect of reestablishing organic matter that made Chandler move into the laboratory, and the publication of disparate sampling results by Rodale and *Acres U.S.A.* was of signal importance. Chandler used the natural method to test the available fraction of humus, not the total carbon, because the undecomposed carbon does not contribute to this or that crop's benefit. It is the decomposed active section of humus that really matters.

Texas Plant & Soil Lab got academia's attention. Commercial labs took notice because they were not including the buildup factor in their recommendations. It may have been happenstance, but conservation tillage became current coin not long after Chandler's revelations could no longer be ignored. The United States Department of Agriculture gave its blessing to appropriate research. Conservation tillage and burn-down with glyphosate promised a gentler role for chemistry, one that may have tiptoed opposing views toward consensus.

In the wake of research by T. Senn of Clemson University and Bob Petit of Texas A&M, the word went forth that a gallon of humic acid — as an extract from humates — delivered astonishing results when put near the seed. The uptake of phosphorus would double and triple. The lesson became transparently obvious once it had been revealed, but it is still not accepted by most traditional soil scientists. The basis of all soils is the humus fraction. It accounts for the life of the soil. It feeds the microbes to release nutrients — for instance, it takes calcium carbonate and acts on it to deliver available calcium.

Organic Matter & Crops

Chandler defends the procedure he uses: "We do the water-soluble fraction and the CO_2 fraction and can do most all extractions used today, so there is very little relationship to the strong extraction methods which show thousands of parts of calcium in the soil which the plant can't get. Plant uptake only calibrates to the CO_2 method when the plant is asked with a plant analysis. It doesn't matter whether the plant is growing in the Rio Grande Valley, the hills of north Louisiana, Russia, the Philippines, Mexico or the upper part of the Midwest with its cold weather to stop the natural method of converting organic matter. Recycling the corn stalk into the next crop becomes more of a challenge. It all goes back to the soil's natural cycle of regenerating the organic matter."

The Rio Grande Valley has the opposite problem. With hot weather in the summer and a mild climate in the winter, the decomposing fraction of the soil hardly ever slows. The rapid conversion of corn stover in the Rio Grande Valley compared to, say, Illinois illustrates the point. Turning in corn stalks and plenty of vegetation with soil moisture and planting two months later works at the Valley's latitude. The same practice at the Chandler farm in central Texas would reveal undigested corn stalks plowed up three years later. Dead soils with little residual fertilizers and little moisture-holding capacity cannot and do not exhibit sufficient microbial activity. Moreover, new varieties of corn do not have the carbon hold that the open-pollinated crop had. Decomposition becomes much more difficult. It takes a healthy microbial population to speed up the process in any case.

It was one thing for the come-lately sponsors of research to claim that the sustainable ag people were unsound, quite another to prove it. The old "no significant difference" saying had to give way to the realizing that "we weren't smart enough to determine the character of the differences," as Chandler's describes the situation.

From Chandler's chair, real progress starts with observation. A recent trip to a pecan grove yielded a bit of knowledge overlooked by even seasoned growers. It is the principle of root flair. At the base of the tree, the root flairs beneath the bark at the beginning of the cambium layer. Howard Garrett, the gardening broadcaster of Dallas, Texas, has found that the root flair has to be aerated. The meaning is clear. In the urban situation, "we have planted trees too deep," Chandler explains. "As they get older, and challenges take their toll, they die." A new tool, the air spade, has come to the market. It blows the soil away to expose the flair of the tree. As soon as the flair of dying trees is exposed, those same trees reverse and come alive, and even more quickly if aided by natural/organic methods and beneficial soil microbes.

Malcolm Beck is putting compost into the equation to install humus in the soil. In the pecan grove that Chandler visited, certain trees were covered with aphids, insects ravaging and culling the weak. Not one of the infested trees had an exposed root flair. As Howard Garrett puts it, "root flairs need air." Garrett has now enlarged his observation to include all trees.

Sick Tree Diagnosis

Soil is the foundation of all life: healthy soils produce healthy plants that sustain healthy animals!

Roots must penetrate the soil. Check for compaction and other problems when soil testing. Roots must have balanced available plant nutrients to sustain healthy growth.

Soil test top and subsoils for available nutrients, salts, and tilth (for water & root penetration). Sample the topsoil (0—12") as well as the subsoil in 1' increments down through 4' or to parent material. Make representative composite samples at each level from four sides of the tree at the drip line (do not composite "mix" different depths of samples). If there are large visual or physical differences, keep each core in separate bags, send with explanations, *let the lab do the mixing of different sides if and when needed.*

Plant trees with a spade or hand trowel, trench outward from the edge of root ball. With a water or air hose expose the root develop-

---

ment to be sure they were planted correctly at the right depth and the roots are extending outward and down from the developing root ball. Often the roots grow inward and must be adjusted for normal growth. Observe for obstructions, walkways or driveways and pavement in the root zone which should extend outward from the trunk up to 1 or 2 times the height of the tree.

Check the Root Flares

For naturally occuring & older planted trees, uncover (gently) the root collar and the root flares of the major roots, between 4—11 in number. The collar is usually a swelling at or near the groundline at the base of the bark and where the main roots begin to flare outward, before entering the soil to a depth 8-12 inches and then sending out feeder roots to find moisture and nutrients. These *main anchor roots must breathe* to support healthy feeder roots. Smothered root flares are the number two cause of declining tree health. The number one cause is trunk damage from weed eaters, mowers and other implements. Crape Myrtles will sprout suckers when root flares are covered!

Expose root flares by scraping with hand trowel, using air pressure or gently washing away soil, mulch or fill, all the while avoiding damage to the tender roots. Do not fill or cover the root area at the base of the tree, especially with mulch or plants, to a distance of 2-4' from the trunk.

Feeder roots support growth with young fine root hairs that compete with sod and other vegetation from the surface down through the aerated soil up to 6". This major feeding zone is concentrated from the drip line back towards the trunk about halfway and then that same distance outward from the drip line. This is the area that needs to be fertilized and covered with mulch especially for young trees.

Disease and insects usually attack stressed and weakened trees (nature's law — survival of the fittest).

Follow soil test recommendations — balance nutrients, especially lime if calcium is needed and should include humus, energy and biological products.

With salt problems soils require internal drainage and solubilized Ca.

After the soil inoculants (biologicals), apply a 2-4" layer of a good livestock-based compost around the plants drip line, starting 2' away from the trunk and extending beyond the drip line to about the same distances as half the way back toward the trunk.

With a water jet on the end of a water hose make small holes 2-4' deep at random many times in the composted area. This moves the nutrients from the compost into the root zone to feed the beneficial microbes in the soil and the F-68 type (live-active) soil inoculants, to where they can fight the pathogens which feed on the roots.

References: Howard Garret (flare expert) DirtDoctor.com, *Malcolm Beck, GardenVille* malcolmbeck.com *& G. Sandy Ross, ASCA* GSRTREE@aol.com, *University of Missouri* conservation.state.mo.us, *Air Spade* air-spade.com.

Testing Soils for Paramagnetism

In the Rio Grande Valley you can go from alkaline to acid, from deep, heavy clay soils to blowing sands. All levels of organic matter assert themselves. During the past two decades, paramagnetic fines of certain rock minerals were recognized and implemented. While Chandler studied soils at Louisiana State University, a fellow named Phil Callahan was picking up his degree in entomology. Later, Callahan published *Tuning in to Nature*, an envelope-pushing study that ratified the idea that insects are a disposal crew, and that they exist as predators to remove crops unfit in the evolutionary scheme of things. Callahan found himself drawn to magnetism and paramagnetism. He developed instruments to test the paramagnetism of materials that made up soils. It was an area of investigation never envisioned by the people who presumed to legislate biology. Callahan's books on insects and magnetism and paramagnetism prompted Chandler, with Malcolm Beck donating an instrument, to collect, test and hold in escrow data for a future database calculated to shock academic agriculture out of its lethargy. The experience was much like the desert traveler who was told to pick up a handful of gravel and not to look at it until daylight, at which point he would be both happy and sorry. The gravel turned out to be diamonds. He was happy to be rich, but sorry that he had only a handful.

The volcanic material that is laced across Texas may well be better than diamonds because it will grow crops, which diamonds won't do. Speaking of variables, nothing is as variable as the paramagnetic values of the soils. A few growers have discovered rocks and dusts that help out-produce fruits and vegetables above the norm that others consider excellent. This production ships better, looks and tastes better, resists early frost, but every case seems somehow shrouded in mystery. Now that a lab in Edinburg, Texas is looking at paramagnetism, the mystery is beginning to clear. A laboratory worker who has been farming for over 50 years signs his reports Conan Wood. He has watched those enclave farmers, tested their soils, and now evaluates paramagnetic qualities of soil. When he sees paramagnetism readings jump into the hundreds, he makes the connection.

Talk to Chandler for an hour or two. He will cite reports of intrepid growers who made a good living, even a fortune by seeing what they look at, by taking pragmatic observations to commercial possibilities, the catalyst being unknown paramagnetism of the soils. Here is a variable few agronomists consider, much less understand. That's the way of progress. Foliar fertilization took hold in the Pacific Northwest fully 20 years before a single university became aware of it. When Iowa researchers finally got around to evaluating the practice, they overlooked the real nature of the foliar route and the stomata intake, and relied on NPK, not on the natural gibberellins and other hormones in seaweed or in the Medina product — cow manure treated with ocean water. The forerunner of half a hundred microbial products, E.E. Pfeiffer, the crusty old promoter of composting, once said it all. "Yes, your salt fertilizers will work, but you have to have compost to make them work."

Lava rock minerals are now being spread on farm ground by the truckloads, and even though some states proscribe soil products not sanctified by law, the road is open and clear, and the most successful growers are attaining the sustainable mode, never to return to unbalanced salts and reckless toxic rescue chemistry.

The thing some few growers observed, and University Extension didn't, was how pockets of overflow and paramagnetic influence favored one geographical location over another. The first of the big Rio Grande dams, Falcon came online between Rio Grande City and Laredo. Eisenhower's Corps of Engineers then built another dam (called Amistad) north of Del Rio. Those dams created the lifeblood for the irrigated Valley.

The Value of Water

Cyclical rains can keep the reservoirs full and make former dry land outperform historically irrigated acres. In the main, it is regulated water that makes the Valley perform. West Texas on the High Plains is the locus of irrigated cotton, pumps usually handling the distribution chore. However, the high cost of energy is causing many growers to long for a gravity system and reduced costs. The cost of pumping and water reserves mandate drip irrigation, especially in the Valley, center pivots being relegated to the uplands.

Still and all, nothing cancels the reality that the soil is the most recognized of variables. "It all comes back to the load in the soil," says Chandler, "the dynamics being to make the fertility available. Too often the industry has gone strictly chemical and that's where cation exchange and soil testing relied on harsh chemicals to determine the availability of nutrients, but we left the plant out of the equation."

The Effects of Nitrogen

Too much nitrogen too early is both the original sin and the continuing transgression of agriculture. As a consequence, vine-stalk production, not fruit, results. $N + H_2O$ = vegetative volume, the cardinal rule of growth. The greater the stalk production, the greater the shedding of lower leaves and less the production of carbohydrates that account for the fruit. Vegetative production is not the part the farmer sells in most crops.

When the seedbed fills out the plant, the lab calls in the petiole test or leaf analysis earlier. This seems to be the key to bringing nitrogen into balance with the minerals. "If we can hold the nitrogen down and get the minerals up, and rely on the phosphate as an indicator of root activity, we can move the plant into the fruiting stage, at which point it needs the nitrogen," explains Chandler.

Texas A&M's Stickler and others have published a paper on the requirements of cotton, wheat, grain sorghum, and corn. It states the requirements by stage of growth: so much for nitrogen, phosphorus, and potash. Thus, it becomes a case of supplying nutrients — and limiting nutrients. It is only when the plant starts fruiting that ample nitrogen is required, along with potash. Phosphorus in small amounts is a rather constant factor, whatever the source.

This insight and wisdom runs counter to the professorial conceit that asks for a season's phosphate even before planting for a harvest months later. By that time, the soil has tied up more phosphate than the plant takes, an observation wasted along with the statistic that less than 15% of the phosphate ends up in the corn, wheat, milo, soybeans, cotton, etc. It takes microbial action to keep phosphorus available, whatever the crop.

Microorganisms & Compost Tea

One of William A. Albrecht's most important papers was "Insoluble, but Available." Modern fertilizer purveyors opt for water-soluble fertilizers. "Not a good idea," says Albrecht. Elaine Ingham, who speaks of the soil foodweb, defers to the microorganism the task of making insoluble nutrients available. This happens to be a consideration that has long separated natural organiculture from the postwar systems called conventional.

Some of the acres under Chandler's tutelage grow pecans. Pecan growers took a serious hit a few years back when DuPont's herbicide Banvel figured in making pecans unsaleable. The USDA organic program at Weslaco has had a pecan plot under bio-correct management for some five years as these lines are being written. Based on a show of hands, it appears that fully 75-80% of the growers who have seen it are ready to embrace natural/organic input methods. Those who watched the procedure became acquainted with Ingham's compost tea, apparently a surefire control for insects, scab and improved fertility.

Organic Pecan Production

A Side-By-Side Organic & Conventional Orchard Comparison

Principles of Organic Production

1. Increase soil organic matter
2. Balance soil nutrients
3. "Balance" soil microbial populations
4. Disease control through improved plant health
5. Insect control through improved plant health
6. Monitor plant health through tissue tests
7. Knowledge of principles of pest control
8. Daily observations
9. Use compost teas

Pecan Orchard

1. Comanche Co., Texas
2. USDA-ARS Organic Experimental Orchard
   a. 5 varieties — 422 trees
   b. 3 rows Desirable — 3 treatments
   c. 4 rows Caddo — 4 treatments
   d. 8 rows Pawnee — 9 treatments
   e. 9 rows Wichita — 10 treatments
   f. 14 rows Cheyenne — 15 treatments
3. Conventionally-managed orchard
   (Desirable, Wichita, Cheyenne)

Experimental Plot

1. 1,000 sq.ft. surrounding each tree
2. Equals a radius of soil amendment coverage of 17.8 feet
3. Soil treatments applied two or three times each year
4. Soil compost tea — three times per year
5. Foliar compost tea — monthly following bloom stage

6. Foliar zinc — one soil treatment and three foliar treatments

7. Control — three 50 lb/acre nitrogen soil treatments

Returns on Sales

1. Using the 2005 data: Control yield is 25.9 lbs./tree. Assuming 35 trees/acre equates to 906 lbs/acre. Sales at $2/lb. equals $1,813 dollars/acre

2. Poultry litter + mycorrhizal fungi yield is 44.1 lbs./tree. Equates to 1,545 lbs./acre. Sales at $2/lb. equals $3,087/acre. Or an increase in returns on sales of $1,274.

3. Value-added sales. However, if the grower sells the organic pecans as Certified Organic (using the December 2007 price of 5 lbs. for $36), sales returns for the poultry litter + mycorrhizal fungi treatment is $11,124/acre or an increase of $9,311/acre.

Table 1. Rainfall records from USDA Office, Hamilton, Texas

Year	Total (inches)
2002	33.87
2003	31.18
2004	49.12
2005	21.25
2006	23.24
2007	43.61
Normal	31.87

Table 2. The 2005 pecan yield (grams per tree) for Pawnee variety pecans from the pecan orchard, Comanche County, Texas

Soil treatment	Yield	Yield	Rank
	lbs./tree	grams/tree	
Poultry litter + mycorrhizal fungi	44.10	20,144 a	1
Poultry litter	38.38	17,409 ab	2
Poultry litter + microbes	37.91	17,196 ab	3
Biological system #1	37.84	17,164 ab	4

Soil treatment	Yield	Yield	Rank
Garrett's organic program	36.48	16,547 abc	5
Compost	33.89	15,372 bcd	6
Biological system #3	29.87	13,549 bcd	7
Biological system #2	27.85	12,632 cd	8
Control	25.85	11,725 d	9

lsd=8.97
cv=29.0

Table 3. The 2005 pecan weight (grams per 100 pecans) for Pawnee variety pecans from the pecan orchard, Comanche County, Texas.

Soil treatment	Yield	Rank
Poultry litter + mycorrhizal fungi	728.2	1
Compost	708.7	2
Garrett's organic program	704.6	3
Poultry litter	698.6	4
Biological system #1	679.8	5
Biological system #2	665.9	6
Control	662.7	7
Poultry litter + microbes	652.2	8
Biological system #3	634.2	9

Table 4. The 2006 pecan yield (grams per tree) for Pawnee variety pecans from the pecan orchard, Comanche County, Texas.

Soil treatment	Yield	Yield	Rank
	lbs./tree	grams/tree	
Compost	3.63	1,645 a	1
Poultry litter	3.10	1,408 a	2
Poultry litter + compost	2.78	1,260 a	3
Poultry litter + mycorrhizae	2.59	1,176 a	4
Poultry litter + rock minerals	2.35	1,066 a	5

Soil treatment	Yield	Yield	Rank
	lbs./tree	grams/tree	
Biological system #1	2.32	1,052 a	6
Kinsey organic balance	1.91	868 a	7
Control	1.89	856 a	8
Poultry litter + microbes	1.62	732 a	9

lsd=2.24
cv=102.3

Table 5. The 2006 pecan weight (grams per 100 pecans) for Pawnee variety pecans from the pecan orchard, Comanche County, Texas.

Soil treatment	Yield grams/100 pecans	Rank
Poultry litter + mycorrhizal fungi	728.2	1
Compost	708.7	2
Poultry litter + compost	704.6	3
Poultry litter	698.6	4
Biological system #1	679.8	5
Poultry litter + rock minerals	665.9	6
Gebert Control	662.7	7
Poultry litter + microbes	652.2	8
Organic balance	634.2	9

Table 6. The 2007 pecan yield (grams per tree) for Pawnee variety pecans from the pecan orchard, Comanche County, Texas.

Soil treatment	Yield	Yield	Rank
	lbs./tree	grams/tree	
Poultry litter + mycorrhizae	45.09	20,452 a	1
Biological system #1	41.71	18,919 ab	2
Poultry litter + compost	38.89	17,640 abc	3
Poultry litter + microbes	37.66	17,082 abc	4
Poultry litter	36.48	16,547 abc	5

Soil treatment	Yield	Yield	Rank
	lbs./tree	grams/tree	
Organic balance	33.79	15,326 bc	6
Control	33.39	15,145 bc	7
Compost	33.22	15,068 bc	8
Poultry litter + rock minerals	29.25	13,267 c	9

lsd=10.15
cv=31.1

Table 7. The 2007 pecan weight (grams per 100 pecans) for Pawnee variety pecans from the pecan orchard, Comanche County, Texas.

Soil treatment	Yield	Rank
Poultry litter + rock minerals	873.6	1
Poultry litter + compost	865.9	2
Control	854.5	3
Poultry litter + mycorrhizal fungi	851.0	4
Organic balance	843.6	5
Poultry litter	842.0	6
Poultry litter + microbes	832.8	7
Compost	830.8	8
Biological system #1	827.9	9

Table 8. Comparison of pecan nut characteristics among varieties for conventionally versus organically grown pecans following five years of organic soil and foliar applications.

Variety	Orchard	Nut Weight	Kernel Weight	% Kernel
			(grams/100 nuts)	
Desirable	Conventional	729.96	395.75	0.54
	Organic	928.35	481.98	0.52
	% increase	27.18	21.79	-3.70

Variety	Orchard	Nut Weight	Kernel Weight	% Kernel
		(grams/100 nuts)		
Wichita	Conventional	519.74	356.44	0.69
	Organic	579.60	392.55	0.68
	% increase	11.52	10.13	-1.45
Cheyenne	Conventional	566.06	326.97	0.58
	Organic	708.69	402.96	0.57
	% increase	25.20	23.24	1.72

Table 9. Soil foodweb results soil samples taken under four treatments following the 2007 harvest.

Measure	Conventional	Poultry liter	Poultry litter + Compost	Control
Active bacteria	44.2	56.4	58.9	44.1
Total bacteria	388.3	698.0	756.0	602.0
Active fungi	16.6	37.1	37.3	23.6
Total fungi	273.0	426.3	349.3	412.6
Flagellate	2,780.0	39,995.0	16,299.0	1,345.0
Protozoa	3,849.0	14,280.0	13,900.0	6,965.0
Ciliates	78.0	385.0	1,177.0	229.0
Nematodes	0.92	1.02	0.97	0.056

Note: Conventional denotes the conventional chemical orchard. Control denotes the treatment using the organic approach except the fertility program is synthetic fertilizers; the soil and leaves were sprayed with compost tea. Statistical analysis was not conducted to insufficient replications (three).

Joe M. Bradford & Larry M. Zibilske, USDA-ARS, Weslaco, Texas

Both a cause and an effect are what happens to the humus fraction of the soil and to the nutrient fraction across the board. Lessons harvested from experience seem to jump from crop to crop. Take peanuts. When the available calcium gets low in the major root-feeding zone, inferior fruit results. Chandler routinely calibrates this back to the soil test — the plant uptake — in terms of quality.

Adding Energy to Soil

Supplementation with a natural sugar like molasses delivers more fruit set and the refinement of management discussed here enables quicker fruit setting. The molasses sugar is a source of energy for the roots and all the microbes that grow in the vicinity of the roots. The absence of sugar requires the plant to deliver carbohydrates to the microbes for their survival. Additionally, humic acid and humate products are a source of food for microbes.

Chandler is forever searching, learning, and teaching. His contacts with Dr. Elaine Ingham of Soil Foodweb, Melinda Kneese, of MK Labs, and Kathleen Draper of AgriEnergy opened his world to in-depth soil biology and compost tea. Ingham is one credentialed scientist who has introduced the microbial connection discussed by George H. Earp-Thomas in the early years of the 20th century. Ingham concentrates on what is missing in the soil, and the roll call proceeds by name: fungi, protozoa, bacteria, genus and species that obey the biblical injunction to increase and multiply, in this case in a compost tea. Here is an art that bows to science the way a lion obeys a trainer. With sufficient food, energy and favorable environmental conditions inoculation via compost tea triggers a microbial explosion.

Weak plants do draw insects and fungal diseases. Excess nitrogen balances itself out. The plant will tell you to bring in carbohydrates to convert nitrate nitrogen to amine nitrogen to aid in fruiting. The plant gets the right form of nitrogen through the biological process. Carbohydrates must be delivered to feed those microorganisms. Leaves reach out for solar revenue to produce those carbohydrates both for the fruit and for the soil. Thus the use of molasses (sugars) on the plant and soil to restore the carbohydrate load.

Carbohydrate Functions in the Plant (Sugars/Molasses)

Added carbohydrate (sugars) increase plant functions for maximum economic production.

Sugarcane molasses is the best natural source — table sugar is more concentrated but also works!

Environmental factors that affect when and how much sugar to use:

      a. How much excess nitrate is in the soil,
          and in plant sap (petiole test).
      b. Soil moisture conditions.
      c. Sunlight intensity.
      e. Wind.
      d. Temperature.
      f. Fruiting stage/load.
      g. Growth/vigor (shade on the lower leaves).

The right supplemental amount of sugar at the right time can improve fruiting and help produce normal plant growth with less attraction of disease and insects.

Natural sugars are needed for healthy plants — fruit production, plant development and maturity.

Roots take nutrients from the soil and transport them up the stalk through the petiole (stem) to the leaves where the sunlight aids the production of photosynthates (sugars are not the only product of photosynthesis) carbohydrates (C, H & O), principally glucose ($C_6H_{12}O_6$) and then many other sugars and photosynthates (hormones, enzymes, etc.) are formed.

Plant sugars and other photosynthates are first translocated from leaves (boron is essential for the translocation) to a fruiting site. If fruit is not available, the sugars, along with excess nitrates, spur the rapid vegetative growth of the plant at the expense of creating fruiting bodies (first site) for the storage of the sugars. Once the proper balance of environmental factors (heat units, light intensity, soil moisture, nutrient balance, etc) are met the fruiting buds form and then fruit formation gets the first crack at the sugar supply.

Any excess sugars are then translocated to the number two site, (growing terminals,) to speed their growth. The leftover sugars, etc., then go to the number three site (the roots) to aid their growth. Here the new root hairs take up nutrients, especially PO_4, to help continue the cycle of sugar and other photosynthates production, for fruiting, growth of terminals and roots.

Added Sugars Can Aid the Plant in Several Ways

• Molasses is probably the best outside sugar source (containing many types of sugars, humics, K, Ca, Fe, etc.).

• Sugar can be added to the soil through irrigation water, by drip, pivot or watering buckets.

• In the soil it can supply energy to microbes feeding on carbon (humus) to stimulate the conversion of nitrates to the more efficient NH_2 form of N. This helps plants synthesize protein directly without them having to utilize their natural sugar for energy for the conversion of NH_4 to NH_3 to NH_2 (the amine form).

• The roots can directly absorb some of the sugars into the sap-stream to supplement the leaf supply to fruit where it is most needed, and *also* directly feed the roots and adjacent microbes for continued productive growth.

• This *added* sugar can also help initiate fruiting buds in a steady-slow fashion while maintaining normal growth (important for melons, peppers, cotton squares, peanuts, etc.)

• *Excessive* amounts of *added sugars* applied by foliar methods can shock the plant resulting in shortened growth internodes, increased leaf maturity and initiation of excess fruiting sites. This can be a short-term effect lasting only a few days. Pollination, soil moisture, nutrient balance and sufficiency as well as adequate light for photosynthate production decide how much of the induced fruit can mature.

• *Added sugars* can be beneficial when nitrates are excessively high in the soil and plants.

• *Excess NO_3 in the soil can be toxic.* The N must be converted to the NH_2 form for the development of protein N for the plant to properly assimilate. The conversion requires energy so the plant's supply of natural carbohydrates (sugars) is utilized at the expense of better fruit development. Adding extra sugars to the soil supplies energy for the soil microbes to convert the nitrate so that the naturally produced sugars in the leaves do not have to be wasted supplying energy for the photosynthesis processes in the leaves, but then can directly support fruit production.

• Also, roots can utilize the extra sugars for their normal growth and plant functions, especially when the leaves are not producing adequate sugars for fruit, root and shoot growth, which causes plant cut out.

• Sugars can be added to the soil in water and fertilizers.

• In the plant, foliar-applied sugars are utilized much faster than soil-applied, and there can be a shock effect if overdosing occurs. Sugars can be directly assimilated into the photosynthesis process occurring in the leaf, speeding maturity and producing more natural sugars. This reaction occurs within hours of application and fades after three to seven days. It supplements the naturally produced sugars and the excess is transported to the fruit producing areas to initiate fruiting buds or supply fruit development. Excess then goes to the growth terminals to sustain new growth and future fruiting sites. The remaining sugars go to the roots to sustain their new growth.

• Cloudy days and low sunlight intensity reduce natural sugar production causing less fruit set or sloughing young fruit, longer space between nodes and fewer fruiting buds.

• Sugar is a source of energy for beneficial soil microbes. Microbes existing on soil organic matter can multiply faster when there is an abundant energy supply. Sugars supply energy for rapid microbial decomposition of raw organic matter and thereby release plant nutrients to roots, and conversion of nitrates to the organic form (amine nitrogen — NH_2) that can be directly and efficiently assimilated into the plant processes.

• Prevent leaf burn from repeated foliar sprays. Carbon (carbohydrates) buffers the caustic effect of many chemical nutrients and pesticides. Sugars, humus compounds, urea and other carbon-containing compounds can protect leaf surfaces from damage and increase efficacy, but one must use proper amounts!

Sugars have an array of functions on the soil foodweb. Too much or too little can greatly affect the plant development of fruiting and/ or maturity!

• This is an oversimplification of very complex biological plant functions based on many published articles and on many trial-and-error crop-log petiole programs on hundreds of fields across thousands of acres.

Cation Exchange Capacity & Soils

Much of agriculture speaks in terms of cation exchange capacity. "This," explains Chandler, "relies on a strong extraction method. Unfortunately, the plant can't run down to the drug store and buy an extraction solution. It has to make do with the acids nature has provided." Chandler and his workers try to duplicate this system with their CO_2 extraction method.

Before natural organiculture captured the public's imagination, few agronomists gave much consideration to the business of microbes preparing the meal for plants. The point is now clear, however, that it is availability, not dead reckoning, that matters. It is always easier to calculate cation exchange capacity or percent of base saturation than it is to calibrate and measure the actual intake of the plant.

The Chandler norm for measuring holds true for any crop. "Rather than recite a theory of availability, we ask the plant," says Chandler, "but first we have to accept Albrecht's theory — the nutrient has to be available." To this end, Chandler studies crops one at a time: sugar cane, cotton, small grains, vegetables, fruits and nuts. They all respond to the same stimuli. Then there's the vegetative volume crops like celery and the multi-foliate crops with their succession of setting fruit as the plant develops: soybeans, peppers, and in a manner of speaking, grapes and pecans. Citrus has to be considered along with the fruit and nut crops. The multiple fruiting crop loads up and matures, all the while setting a succession of fruit. It opens in different stages, not at one time as is the case with citrus or pecans. Much the same is true for almonds and other nut crops. "It is generally considered in terms of plant analysis once a year," Chandler explains. "But the requirement of that crop as it sets a fruit is entirely different from what it needs when the fruit initiates growth or passes on to maturity."

Visual Crop Checking

A testing procedure of crop-logging was developed by the sugarcane people. Dr. H. F. Clements' *Sugarcane Crop Logging and Crop Control: Principles and Practices* is an excellent publication that ties a total production program together. Later, Al Lengyel used the process for citrus plants, analyzing old and new leaves monthly to establish physiological needs at all growth stages. It provided a valid picture of all stages, not just the end stage in the production cycle. The picture changes as the crop sets. When its pattern is held, a tree will set more fruit than it can hold. This assessment took root as the new art developed. Thus the reason for studying the excess. If the nutrient level is adequate, it will hold its fruit. If more fruit is held, then feeding must proceed at a higher level in order to mature that fruit or the end-of-the-season fruit will be imperfect, often unsaleable. A living balance thus becomes the objective.

Noted Soil Biology Events

Smart or not, there have been two great leaps forward in soil biology during the lifetime of Esper K. Chandler. One was protocol developed by William A. Albrecht and associates, namely the ratios for the cation exchange equation and the anion requirements of the crop. It became the benchmark for an era even though it ushered in a problem all its own. It was brilliant, frustrating and universally applicable. It invited a chorus of question marks.

To an old cotton man from the hills of northern Louisiana, it seemed impossible to calibrate the growing crop to the theoretical equation. For this very reason Chandler returned to the Arizona extraction method of Albin D. Lengyel, the second great leap forward. How does the plant take the nutrient out of the soil? It does it by secreting carbon dioxide into the soil moisture, thereby creating a weak carbonic acid. Albrecht was dealing with stronger soil extractions. Chandler found that when he measured the differences in the soil, the carbon dioxide or natural method proved much more definitive because the soil variation had to be accounted for.

Chandler and Lengyel's way involved scrutiny of this crop, this plant, this petiole, not as a sweeping generalization for the season, but for this day with this temperature and this amount of sunlight. It became a devastating adjustment in the state of knowledge, and it helped kick open the door for more serious consideration of the premises that Sir Albert Howard and J.I. Rodale had exposed all along.

When Chandler assumed control of the soil testing laboratory in 1980, his first objective was to calibrate the uptake of nutrients to the petiole test. The art was all but lost. The University of Arizona still had an official method for carbon dioxide extraction, which had once been used on salty-alkaline soils west of central Kansas. As with many of the treasured lessons from the post-war chemical era, that art and science had become eclipsed. The genius of Chandler and Lengyel was ratified when Chandler moved the method to the forefront in meeting the challenges brought on by "shooting from the hip" agronomy that had failed to consider the long term. As with Nature herself, the Chandler system is not subject to full automation. For this reason, Texas Plant & Soil Lab is probably the only one still using the CO_2 method. Florida was probably the last to officially abandon the method citing it as too time consuming and expensive while acknowledging its benefits.

Chandler defends the CO_2 method. The cation exchange is based chiefly on texture: the more clay, the higher the exchange. This is modified and improved by increasing the organic matter, the available humus fraction. Failure to take into account the undecomposed portion of organic matter imposes a variable that Chandler figures ought to be extinguished. "If you use the active humus extraction, you get a better deliberation than the cation exchange capacity, which does not enable calibration of uptake by the plant root."

The suggestion here is that the plant root does not care what kind of soil it is growing in: alkaline, salty, highly loaded, well-leached, arctic, tropical or temperate. The root does the extraction by the same chemical method. This claim has been validated by a database computed from soils the world over. "We still get a better calibration by asking the plant via plant analysis what can be gotten out of the soil." This much stated, Chandler has to admit that, "on a logarithmic basis, the more we think we know, the lower on the scale we go." He goes on to amplify . . . "The soil we all depend on is so complex, so dynamic, so constantly changing, that by the time we think we know something, we find that we're digging a hole deeper rather than climbing a mountain." There are no absolutes in nature, just guiding principles.

Making Choices

At almost any agriculture conference, commercial booths are seen as an index of the intellectual feast that sustainable farming has become. Half a century ago, a few daring innovators such as old man Martin fermented ocean water with manure from lactating cows to create the precursor to Medina. These preparations, dozens and hundreds of them, now challenge the industrial absconders, and serious agronomists like Chandler are busily testing and validating the products of the innovators, one by one.

Faced with a plethora of tests, Chandler somehow escapes the shortfall of being too laboratory-oriented. In fact, he spends a great deal of time walking the soil, getting its feel, enjoying its smell. These practices instill a practical approach that tells him when the tests are telling both what they know and what they expect a short way down the track.

The purist will warn against sulfur because it can be toxic, yet the crop producer knows that sulfur is absolutely necessary for blueberry production and salty soil remediation. Calcium carbonate is often used, but what does the grower do when the soil has no calcium? Bordeaux mixture is copper sulfate and lime. Is it a nutrient or a poison for mold and fungus? Those who think in terms of poison believe it is a poison. Sustainable growers know that it is a nutrient. Bordeaux mixture makes the plant tissue inhospitable to the invaders. They perish as the plant is led back to health, the copper being a nutrient, and not a toxicant. When soils sink to a pH of four and three, one can doubt the presence of any calcium whatsoever, since it has been mined out with salt fertilizers.

Chandler appreciates a good database more than most, but in a world where one has to earn one's way with services that deliver promptly, few field agronomists have the wherewithal to become bookkeepers. "I can show you the acres and the improvements, and the farmer can show you his bottom line. But the database will come as soon as academia stops shunning natural/organic culture just because they don't like the word 'organic.'"

Thus we have arrived at a situation wherein research is paid for by industry. Accordingly, the results have to comply with the expected, otherwise the results are not published. Often the papers are promptly shredded or ignored. Academia loses respect while hiding behind the randomize, replicate, repeat to obtain statically significant results worthy of peer-review publication syndrome to avoid acknowledging overwhelming antidotal statistics as the top producers add to their equity while being progressive innovators.

Chandler, in all humility, notes that he expects to learn something new each day.

The New Paradigm

New Traditions?

A story is told about a college fraternity that published a pamphlet detailing its new traditions. The tag line said, "These traditions go into effect Monday morning."

Something like this must have happened when the legislated use of NPK fertilizers acquired conventional status in the decade after World War II, sweeping aside traditions going back centuries. It was an exhilarating ride as crop yield records tumbled and new ones were heralded as science triumphant. The decline of organic matter and the poor nutritional quality of crops that needed so much chemical protection went generally unidentified. A few far-seeing agronomists postulated a new paradigm, and one of them was Esper K. Chandler.

The starting point for the new paradigm was a chemical agriculture that conquered all with — as Sir Albert Howard put it — "partial and unbalanced fertilization and toxic rescue chemistry." As a student, farmer, industrialist, consultant, field agronomist, and finally as a laboratory entrepreneur, Chandler never lost sight of the fact that much was wrong with the quick fix. Without a clear understanding of the long-term ramifications of that chemical fix, the industry went blithely on its way pooh-poohing dissenting voices, calling them talented amateurs at best, fertilizer quacks at worst.

"We weren't too much concerned about the soil because we were still in the age of mining our greatest natural resource. We were not paying attention to the regeneration expected of us as permission for life,"

Chandler points out. "The long term was set aside for what can we get from the soil right now." Indeed, some of the farm press — tongue in cheek — suggested a depletion allowance for farmers who mined the soil. The habit of mining the soil continues to this day.

The destruction of farm soil does not always take place under a crop, however. Some of our best soils are sequestered for the public domain, reserved for highways and urban sprawl. Cities expand into suburban sprawl, and builders pave over the best acres. Acres planted in citrus for almost a century are taken out of production in the Rio Grande Valley every day as McAllen and its neighboring towns make the valley one of the fastest growing metro areas south of the Mason-Dixon line.

As a consequence, "we are going to our less-productive soils," warns Chandler. "So now we are starting to wake up with the objective of regenerating the soils based on sustainable methods." The heart of this drive is to rebuild our humus and organic matter, "which feed the biological cycles." This process accelerates the soil's potential for mineral release, much as if barren rocks were making a gift to vegetation.

The new paradigm recognizes the microorganisms that constructed soil tilth in the first place. Chandler sees a huge awakening going on, much of it on the part of people who reject the tradition that went into effect about 50-odd years ago. This is happening over "the objection of scientific agriculture," Chandler asserts. This scene was under the auspices of regulatory officials and trade organizations firmly in the grip of fossil-fuel companies too often tailoring advice to sell the inventoried product in the bin, not to the requirements of the soils.

"When I was hired by the National Plant Food Institute (NPFI), led at that time by Russell Coleman and Sam Tisdale, both made the case for soil testing," Chandler recalls. "Coleman was from Mississippi, Tisdale from Alabama and North Carolina. Tisdale was a leading authority on soil fertility and soil testing, as well as a textbook writer. These men were on the cutting edge of soil improvement, but they had been eclipsed and their call for mineral replacement wasn't heard. The chemical age had proved to be that overpowering!" The job with NPFI exposed Chandler to soil fertility and plant nutrition experts in many states. Rock phosphate was a fine additive for soil building, but fertilizer trends relegated simple rock to the underworld of natural/organics. Chandler's job was to work with southern and southwestern universities, extension services, private industry, lending agencies, and the news media to advance the benefits of soil productivity. The goal was a low unit cost of production. It was understood that a lot of research was being accomplished by good old boys who did not have enough degrees and couldn't get their findings published. This knowledge sat in filing cabinets, and "my job was to seek these people out when it became obvious that they were progressive in their productivity quests," Chandler says. "My job was to use seed money to insert that information into the mainstream, often to the chagrin of so-called settled science."

In those days Bob Pettit, a Texas A&M researcher, worked with humic substances. T. Senn at Clemson University was working with both humic substances and seaweeds. Neither achieved acclaim. As a matter of fact, both were frowned upon because they looked back at the basics for soil fertility, not ahead to more high analysis and fossil fuel company sales disguised as high science.

"I ran into a lot of people who had chosen phosphate work, potash work, etc. We weren't worrying about potash because the cation exchange method revealed lots of potash in the soil. In a manner of speaking, the new paradigm has morphed out of the dominant culture in an obscure manner. In the halfway house of transition, we were using a quick tissue test. It was a precursor to 'ask the plant.'"

Chandler credits Tisdale with midwifing the practice of questioning plants. He bought a tissue test kit for Chandler, and ordered him to spend time with Dr. Nevin D. Morgan, agronomist of the American Potash Institute, a practitioner in several states. Today, Chandler recalls how he would run into workers at Auburn in Tennessee, at Mississippi State, and at other Experimental Stations who were researching legumes and other crops soon to pick up the label sustainable. The needs of crops for calcium, potash, magnesium and micronutrients based on the Albrecht approach were discerned, and information was traded. The laboratory audits told of an ample supply, yet the leaf tests told them the plant was starving. So how was the phosphate to be supplied?

The petiole test told Chandler of a phosphate deficiency in the face of non-supporting anion readings. The conventional wisdom said to rely on high-analysis phosphate. It represented the lowest cost in terms of shipping, but the decision went back to the people who were doing the work that revealed the need. They were academic outcasts who paid no attention to the current consensus.

The Bottom Line

Sir Albert Howard and J.I. Rodale offered a modified code for the farmer that failed to meet the quick fix yearnings of the industry, but their thinking has endured. Fifty years later, it has acquired the respectability of a stable paradigm. Nevertheless, many farmers still believe they have to get it now because there's no tomorrow. Chandler calls it the Harvard Business School mentality, using the nation's most famous business school as shorthand for an obsession with the short-term bottom line mentality that afflicts business colleges and business practice all over America.

"If we do not back up what produces the bottom line, we're back to instant gratification," Chandler says. Least important in this equation is tomorrow and tomorrow after that. "That's what happened to agriculture, yet agriculture is the basis of life. If we don't produce food, fiber and energy, how are we going to sustain a prosperous society?"

The bottom line can't be expressed in terms of dollars. It has to do with making the natural nitrogen and carbon cycles work, and it all depends on the minerals extracted from the soil by plants. It involves promoting a wide range of microbial action, not just fixing nitrogen from legumes. "We have to grow the microbes," says Chandler, "and this is where compost tea is coming on strong. We're growing microbes rapidly to fix nitrogen, among other things. Microbes are made up of proteins."

Innovation with Foliar Feeding

The emphasis on stats and statisticians should have warned those who entered the heavily promoted industrial agriculture that appeared in the postwar era. Instead of ordering the experimenters into the fields with instructions to question the farmers and laborers regarding quality, or taking up a few acres of their own land, the investigators confined themselves to their campuses and devoted their time to bringing results in line with statistics. A thought from the notebook of Sir Albert Howard should have been invoked. Confronted with ineptitude, he asked scientists and administrators whether they held themselves above the farmers and the husbandry practitioners. It was almost certainly a rhetorical question.

It is a matter of record that those who fought back — Chandler included — were either dismissed, ignored, or consigned to the underworld of agronomy. Some of the pioneers of natural/organics knew they were not wrong but could not explain why they were right. Their opponents were quick to take advantage of incoherence or hesitation. And the more things change, the more some things stay the same. The Hudson Institute, a think thank that shills for industrial agriculture, continues its work here in the 21st century, selectively citing only the negatives of organics while mangling true science.

Valid technologies for foliar feeding existed as early as 1951. They were used widely in the Northwest for years. Many years later, Michigan State finally researched the topic, documenting it in a film called *The Non-Root Feeding of Plants*. This would not have happened without aggressive field agronomists shaming academia into it.

The point of this yarn is to illustrate how far Chandler and associates have moved ahead of supposedly settled science. Often Chandler tells his audience — sometimes only a farmer or two — about humic acids and humates with references to non-NPK growth stimulation that brought postwar industrial agriculturalists up fighting from behind their desks. People who gave serious consideration to auxins, cytokinins, and gibberellins, as well as biologicals, caught an equal amount of flak.

It has always been Chandler's position that everything he installs on a working farm can be reduced to an acceptable paper for the referred journals. "But who's got the time?" he says. "Yes, who's got the time when the client farmer needs instant answers during this busy growing season?" As the farm orthodoxy prescribed "more is better," Chandler proved that he could grow more quality bins and bushels with more phosphate uptake, enough in fact to require the rewriting of certain textbooks.

Chandler often wonders how observers could be so dense as not to comprehend the value of Medina. But Chandler took note as he inched ever closer to that academically forbidden "organiculture," urged on by Texas A&M research professors Flake Fisher, Bill Trogdon, and soil test chemist William F. Bennett. Some of the unique crop uses of hormones can be found in the chapter on specialty crops. For now it is enough to examine the survival mode that has come to guide the agriculture not characterized as industrial.

Some 12 years after Michigan State's Sylvan Wittwer made *The Non-Root Feeding of Plants*, the University of Iowa's John Hanway proposed the use of NPK foliar leaf fertilization on soybeans. A multi-state program sponsored by Tennessee Valley Authority (TVA) and the fertilizer industries was a dismal disaster. Results were either marginal or a failure, depending on which press release you read. A finding emerged soon enough that the foliar feeding of soybeans that year failed because the physiology of the plant was ignored by applying foliar feed in the heat of the day. It should be fed in the late afternoon through early morning when the plants' stomata are inhaling. The universities and the industry seemed locked into the idea that gross amounts of fertilizers were required, regardless of the route. That such applications failed to answer plant requirements became an elusive concept, albeit one a few agronomists picked up as they tiptoed into a paradigm often considered off base.

When Chandler took on the task of quality production of political crops and the many specialties then identified with the Rio Grande Valley, he questioned the holy writ of soil fertilization. Chandler, it turned out, was one of the first to point out that correct fertilization of crops called for knowledge of many other things. What is a plant's requirement at each point of the growth cycle? True, a few growers of strawberries, apples, pears, cherries, citrus, flowers for market, etc. started answering growth-cycle requirements with seaweed extracts, yet few producers of the storable commodities — wheat, corn, soybeans, grain sorghum, oats, rye, even canola — picked up the knowledge, in fact they even spurned it.

Chandler's adoption of humic acids via foliar- and drip-irrigation to achieve superb phosphate uptake might have earned him one of those agricultural prizes, but the rules reserve such recognition for people who, as Eisenhower put it, farm with a pencil on a polished desk a hundred miles from a corn field.

The successes rolled up in the Rio Grande Valley during Chandler's early laboratory years were spectacular, and often rejected. For one thing, scientist-agronomist-farmer Chandler realized the folly of many advanced refinements when soils were complexed and confused beyond reason. But with attention to soil balance, efficiency of the new agronomy could be increased 100 to 1,000%. The discovery of nutrient refinement made it possible to help control insect problems. As with cattle, infectious diseases did not seem so infectious when the potential victim buffered assault with immunity — a "physical terrain," as Antoine Beauchamp, a contemporary of Louis Pasteur, put it. Withal, it was hands-on agronomy and specialty crops that highlighted the relationship between the sun's subatomic particles and the earth's trace minerals. Finally, growers learned to seek out natural soil testing and sound nutrition programs.

Embracing New Products

The micronutrients and trace minerals found in ocean water present an awesome roster. The Association of American Plant Food Control Officials (AAPFCO) discontinued a relatively new category addressing specific uses for micronutrients a few years back. Apparently AAPFCO were not yet ready to acknowledge these nutrients and let them into the mainstream, but once opened, the door cannot be closed.

"There is a mentality out there that the bureau person knows better than the grower," says Chandler. "The public is not supposed to have sense enough to make a decision on their own." Rather than sticking to the laws and enforcing guarantees of contents listed on the labels, these bureaucrats want to impose use controls by arbitary decree. More and more, people are waking up to the fact that we have to have free thought out there. The only real answer is on your farm under your conditions — leave a check plot." The trade does not fully recognize products such as Medina, for example, that leave loading docks all over the country ever since Texas A&M failed to suppress it. These trends are expanding explosively. Chandler can name 30 to 50 products that reach beyond NPK to harness the miracle of microorganisms.

Chandler listens to innovators who come up with new products. On a local basis, growers and entrepreneurs have developed their microbial products, some with labels, some merely for personal use. One offshoot of this development is a line of microscopic bacteria that eat oil slicks. It was developed by Carl Oppenheimer, another Texan. Oppenheimer also worked on the original Medina product.

Chandler is optimistic as he reflects on these events. "Much of the new paradigm development is coming from the city dweller. It is seated in horticulture, in the greenhouse, in the garden. Wives and young professionals are concerned about the health of their families. They are fully cognizant as to what chemicals in, on, and around the food supply are doing. They know about the nutritional defects of commodity foods. The query traffic to Chandler often has one urban line, 'I want to grow an organic garden. How do I start?' This person wants nutritious food. The biggest unidentified trend in America is urbanites buying a small ranchette, one big enough for that garden and a few animals. That's where this emphasis on regeneration is also coming from." Many seek out Chandler as a starting point in their quest to take an in-depth top and subsoil test. The overflow has already penetrated mainline agriculture. High costs threaten the industrial model. There is a serendipitous consequence to what is happening: it is the old Albrecht principle that the anatomy of weed and insect control is seated in fertility management, and not in finding a more powerful poison in the devil's pantry.

The world of elemental chemicals has come full circle. That's what the biological revolution is all about.

The Pitfalls of a Toxic Agriculture — The 29th Day

No reasonable person can expect our soil systems to be managed another 50 years as they have been managed for the last 50 years — not when deserts expand, subsoil replaces topsoil, aquifers go dry, oceans become polluted, the air poisons our cells, and produce from the land becomes empty of nutrition.

We are told that French schoolchildren are offered a little parable in order to teach them the futility of compound interest and growth based on debt alone. If a small pond is filled with leaves — one leaf the first day, two the next, four the next, and so on until the pond is full in 30 days, then when is the pond half full? The answer is the 29th day. Technology-wise and economy-wise, we are at the 29th day. What cannot continue, will not!

"It is our responsibility to examine nature's requirements, to discard gods that fail," Chandler says, "and surely the god called toxic technology has failed. The cycles of history assure us of a new era, but they require us to construct it ourselves."

"Where does it come from? Where does it go?" children ask. "You'll understand when you're older," grownups answer, especially when they do not understand the answers themselves.

Reading
the Plant & Soil

The first step on the road to achieving healthy soils able to sustain productive plants is the soil or plant analysis test. For optimum results, the initial test relies heavily on proper sampling (described in Chapter 5 — The Laboratory Asks the Plant). Quality samples submitted to the laboratory and excellent testing methods can produce the most accurate results possible but without an interpretation of the nutrient recommendations that speaks to the grower — all may be for naught. Graphs and charts filled with color-coded lists of numbers speak volumes to those that know how to read them. But the uninitiated may glean just a fraction of the total message. A soil or plant analysis test from a quality laboratory contains much more than just the raw data. Using an integration of field and cropping history with the test results, interpretations and recommendations are formulated to tell the grower the meaning behind the numbers. It is these soil and plant test interpretations and recommendations that matter most and have the greatest benefit for many people.

Chandler was asked to look at several plant and soil analysis tests from different crops and give his expert interpretation of the results. The results follow in this chapter. For each example, Chandler's comments offer new insight and enlightenment about what the results said to him. The soil and plant test samples presented in this chapter are actual real-life examples included here with Chandler's dictated interpretation and recommendations — presented so others can gain a deeper insight into the important messages held within. For each example, the most important messages have been highlighted and explained by Chandler.

Low Micronutrients, High Nitrate & Phosphate Levels

	PPM		% K	% Na	% Ca	% Mg
	NO₃	PO₄				
1	23120	3354	9.15	0.12	2.07	0.34
2	13520	3528	8.55	0.18	1.25	0.39
3	14760	3724	10.58	0.22	1.21	0.57
4	6520	3904	8.86	0.08	1.57	0.41
5		382	7.79	0.09	1.95	0.44
6	7560		7.85	0.18	2.06	0.37
7	4520	3213	6.78	0.23	2.2	0.32
8	2956	3570	6.12	0.17	1.41	0.44
9	7160	4506	6.79	0.17	1.21	0.42
10	13120	3801	8.39	0.1	1.41	0.45
11	8560	5084	8.06	0.08	1.21	0.46
12	10880	3745	7.96	0.14	1.62	0.56
13	7040	3429	6.65	0.09	1.36	0.4
14	12360	4759	8.05	0.07	1.22	0.46
15			8.8	0.12	0.97	0.46
16	21640	2912	6.3	0.13	2.2	0.43
17	13360	2137	8.3	0.19	1.43	0.44
18	16360	3861	10.15	0.25	1.51	0.68
19	10600	3333	10.26	0.1	1.71	0.5
20	1068	3438	6.97	0.1	1.71	0.41
21	0	2363	6.27	0.24	2.33	0.54
22	5440		4.8	0.31	2.36	0.47
23	120	4139	6.15	0.18	1.23	0.44
24	5080	4057	6.03	0.19	1.58	0.5

Best Field — rows 1–15
Average Field — rows 16–24

The nitrogen (N) started very high, and was brought down to a reasonable level after one month.

In the "best field" the phosphate level was much higher than the "average field." This indicates the amount of root activity each week.

Zn	Fe	Mn	Cu	B	Sample	Date
PARTS PER MILLION - PPM						
22	186	23	9	17	TANB052	5/8/08
23	151	20	7	18		5/19/08
65	190	22	16	28		5/28/08
40	127	25	6	13		6/4/08
37	81	20	6	24		6/12/08
						6/18/08
31	120	23	6	18		6/25/08
30	154	24	7	43		7/1/08
16	109	22	8	15		7/8/08
36	67	14	8	27		7/16/08
36	79	13	14	44		7/22/08
48	138	22	11	48		7/30/08
34	115	0.24	11	49		8/5/08
47	92	23	21	29		8/13/08
						8/24/08
32	180	24	9	14	MEX030	5/8/08
21	205	19	7	21		5/19/08
52	237	28	14	29		5/28/08
51	131	29	8	17		6/4/08
54	106	26	10	30		6/12/08
21	100	12	3	24		6/18/08
23	143	22	7	27		6/25/08
30	151	25	8	38		7/1/08
21	112	21	8	23		7/8/08

Overview: These two watermelon fields produced over 100,000 lbs./acre of saleable crop under a drip-irrigation system. A weekly petiole analysis was done.

Recommendations: From the plant analysis, the grower realized he needed to keep the phosphates up and limit the amount of N. He reduced the N applied (usually under 10 lbs./acre of actual N) so there wasn't any excess. Phosphate levels were kept high to get good root activity and maximize yields. He changed to a foliar application of micronutrients instead of an injection method.

High Nitrate
& Low Phosphate Levels

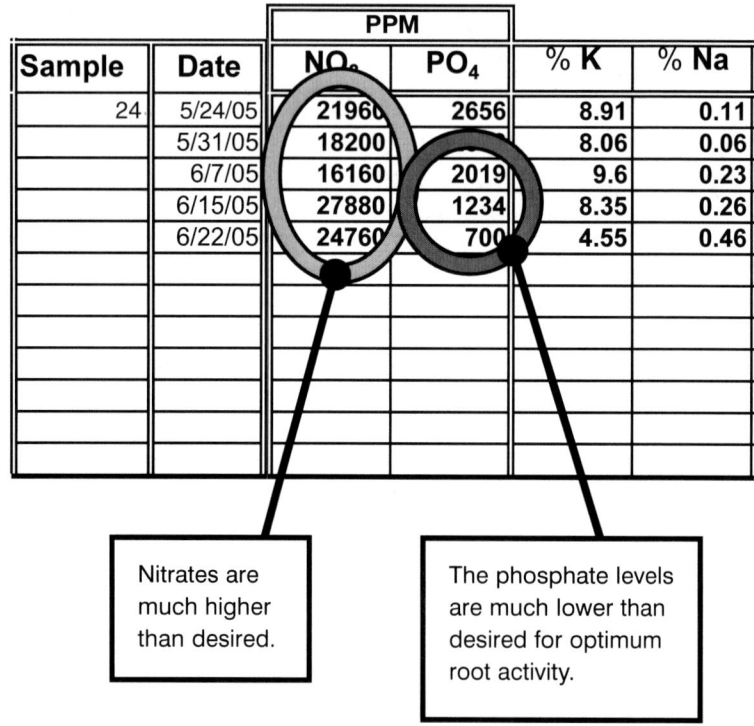

		PPM			
Sample	Date	NO₃	PO₄	% K	% Na
24	5/24/05	21960	2656	8.91	0.11
	5/31/05	18200		8.06	0.06
	6/7/05	16160	2019	9.6	0.23
	6/15/05	27880	1234	8.35	0.26
	6/22/05	24760	700	4.55	0.46

Nitrates are much higher than desired.

The phosphate levels are much lower than desired for optimum root activity.

	PARTS PER MILLION - PPM				
% Ca	% Mg	Zn	Fe	Mn	Cu	B
3.22	0.55	29	110	16	10	60
2.64	0.57	47	215	30	7	49
2.47	0.52	29	172	21	7	50
4.63	0.93	26	76	26	7	53
4.29	0.54	13	115	43	7	45

Overview: This is an average yielding melon field with a harvest of 30,000 lbs./ acre. The high nitrates made this a short-lived field with less than maximum yields. The phosphate level, indicating root activity, was very difficult to keep at an acceptable level. Calcium and magnesium are applied through drip-irrigation.

Recommendations: The key was to keep the nitrogen down to the desired level and build the phosphates as high as possible through the entire season so that the roots are getting more of the balanced nutrients needed to feed the plant.

Low Nitrate
& High Phosphate Levels

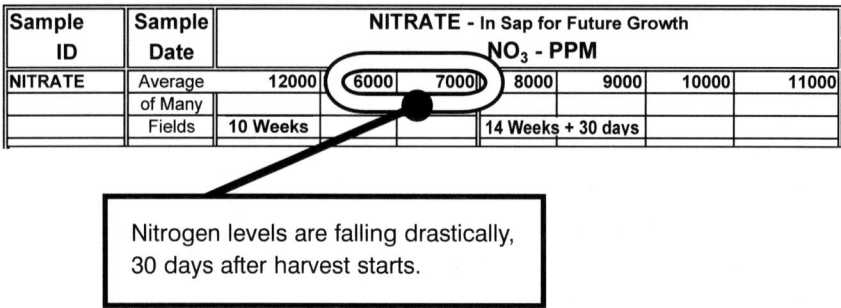

| Sample ID | Sample Date | NITRATE - In Sap for Future Growth | | | | | | |
		NO$_3$ - PPM						
NITRATE	Average of Many Fields	12000	6000	7000	8000	9000	10000	11000
		10 Weeks			14 Weeks + 30 days			

Nitrogen levels are falling drastically, 30 days after harvest starts.

Overview: This watermelon crop was grown on 17 fields in three different farms. Drip-irrigation was used. This past year, the yield was over 100,000 lbs. on 120 acres. Previously the yield was steady at 90,000 lbs. The petioles were tested every week.

Recommendations: The petiole analysis above shows phosphate levels. Higher phosphate levels show that roots are active because nutrients and water are taken up by the young root hairs near the growing tip. Anything that stops the tip from growing such as cultivation, blight, hardpans, too wet, too dry, etc. will slow the uptake of phosphate. For multi-fruiting crops, such as watermelons, the longer they produce, the higher the yields, the greater the profit and lower the unit cost of production.

   After 30 days of harvest (at 10 weeks), root activity was good, but nitrogen levels were falling dramatically. The petiole indicated that the roots were cutting out because the phosphate was down past the threshold. Phosphate was injected on a portion of the field and an increase in phosphorus uptake indicated root activity. At the same time, in a comparison zone of the drip-irrigation system, humic acid was administered. About the same increase in phosphorus uptake occurred. When humic acid and the phosphate were added together, the synergistic effect more than doubled the uptake of phosphorus. The addition of a mixture of four plant hormones (auxins, cytokinin and gibberellin) pushed the watermelons into high yields. Later, multiple hormones, a little bit of phosphorus, multiple sources of biology (rotated) and energy (molasses or fish) were used for even higher yields.

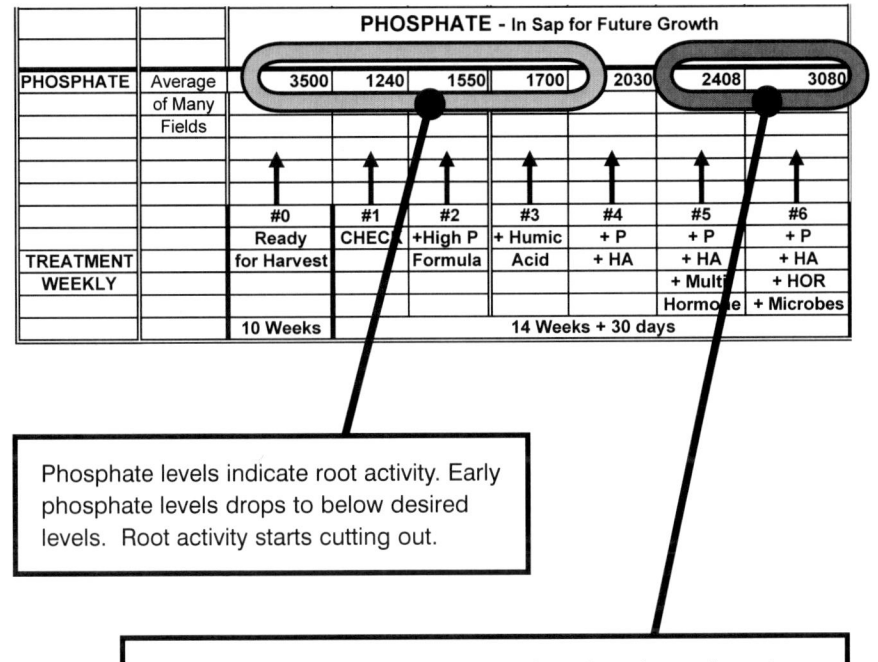

Phosphate levels indicate root activity. Early phosphate levels drops to below desired levels. Root activity starts cutting out.

After the application of humic acid, high phosphorus formula and multi-hormone activators, phosphate levels (and root activity) has increased.

Low Potassium, High Calcium & Magnesium Levels

Field		Text.	O.M.	CO$_3$	pH	Salts E.C.	lb per ac NO$_3$	P$_2$O$_5$
1	YF 200 0-6	5	0.55	O	6.3	0.48	52	2
2		3+	0.95	M	7.2	0.67	59	2
3		3+	0.80	TR	6.4	0.67	23	13
4		3	0.85	L	6.9	0.48	16	38
5	YF 200 6-18	5+		O	6.3	0.48	35	2
6		3+		L+	7.1	0.58	35	2
7		3+		TR	6.9	0.67	9	9
8		5-		VL	7.0	0.48	3	29

Overview: This is a topsoil and subsoil sampling from a 200-acre field of low-yielding coastal Bermudagrass hay under center pivot irrigation. No matter how much fertilizer was applied, the yield was very limited (around 5 tons/acre). When more than ½ inch of rain or irrigation fell, it would run out of the field because of the packed soil surface.

Recommendations: An increase in humus and balancing of the pH, Ca, Mg, P and K were needed to build up the biologicals (and help make uptake of nutrients more available to the plant). Potash and magnesium were increased slightly in the topsoil from a "hi-mag" lime application. After two years of applying very high rates of humus, biology, phosphorus and potassium — yields were tripled over the previous year, soil texture and soil tilth was much improved, and water use reduced by 40%. At the end of the second season, it was easy to dig down 4" with bare hands and find earthworms. The key to success was slowly rebuilding the topsoil's phosphorus and potassium levels in the top 6" without mining the subsoil of these nutrients. The result was to build Bermudagrass root reserves while still taking off impressively high yields.

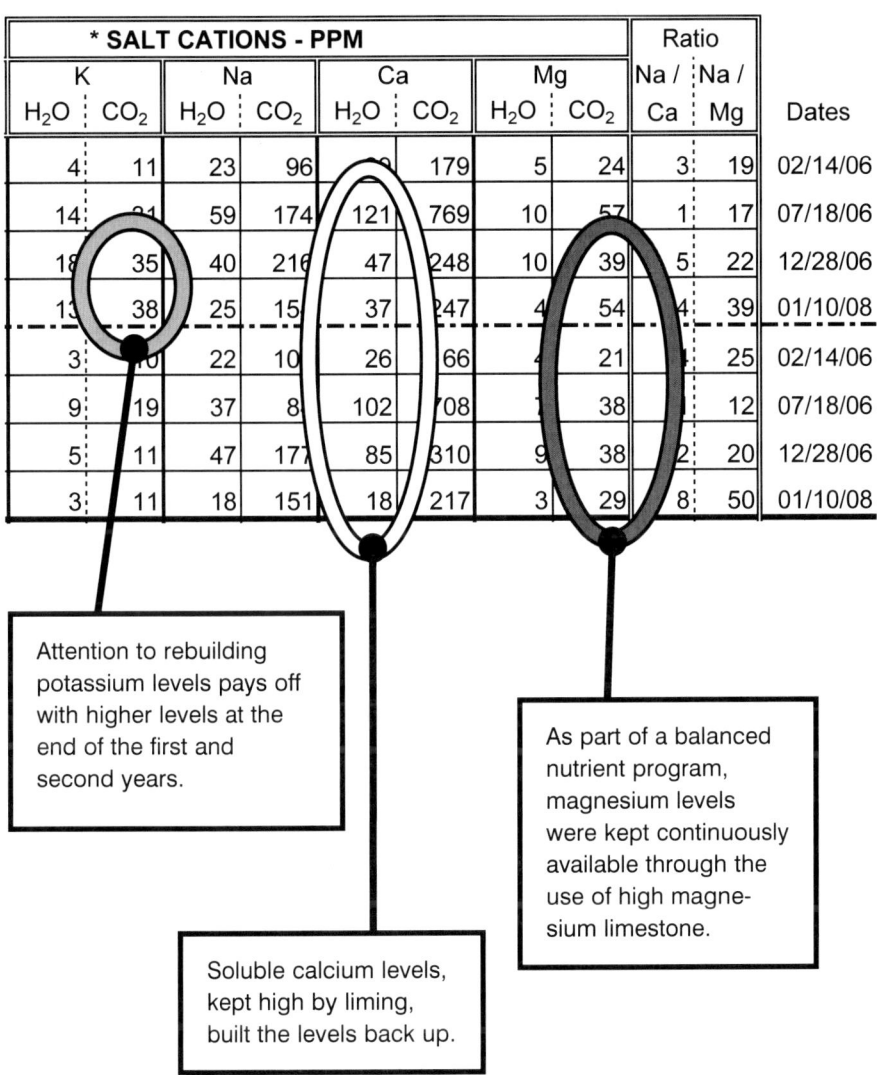

* SALT CATIONS - PPM								Ratio		
K		Na		Ca		Mg		Na /	Na /	
H_2O	CO_2	H_2O	CO_2	H_2O	CO_2	H_2O	CO_2	Ca	Mg	Dates
4	11	23	96		179	5	24	3	19	02/14/06
14	21	59	174	121	769	10	57	1	17	07/18/06
18	35	40	216	47	248	10	39	5	22	12/28/06
13	38	25	15	37	247	4	54	4	39	01/10/08
3	0	22	10	26	66	4	21	1	25	02/14/06
9	19	37	8	102	708	1	38	1	12	07/18/06
5	11	47	177	85	310	9	38	2	20	12/28/06
3	11	18	151	18	217	3	29	8	50	01/10/08

Attention to rebuilding potassium levels pays off with higher levels at the end of the first and second years.

As part of a balanced nutrient program, magnesium levels were kept continuously available through the use of high magnesium limestone.

Soluble calcium levels, kept high by liming, built the levels back up.

Hardpan Issues, High Salt & Calcium Levels

	Field	Text.	O.M.	CO$_3$	pH	Salts E.C.	lb per ac NO$_3$	P$_2$O$_5$
1	North 1'	3	0.35	0	6.8	0.77	74	99
2	" - 2'	4+		H	7.5	0.67	12	23
3	" - 3'	5-		VH	7.5	0.77	16	9
4	" - 4'	5-		VH	7.6	1.01	15	7
5	South - 1'	5-	0.30	0	6.5	2.53	25	61
6	" - 2'	5+		tr	7.1	3.12	20	4
7	" - 3'	5+		H	7.4	2.06	32	6
8	" - 4'	5		H+	7.5	1.73	31	10

The salt level (shown as E.C.) is higher in the subsoil in the south section due to leaching of soluble salts.

Overview: This sample is from the north and south sections of a citrus grove at 1', 2', 3' and 4' depths. A detailed salt analysis (testing N, P, Na, Ca and Mg) was done after several deep flushing rains and a history of regular sulfur use. As a result of leaching in the north section, some phosphorus and sodium accumulated in the subsoil.

Recommendations: The north samples show much better results than the south section. In general, the topsoil doesn't show much difference. The subsoil though, has very high total soluble salts (especially sodium, calcium and magnesium) in the south sample. To alleviate the high sodium, soluble calcium was needed. The calcium can be solubized with sulfur, humus, biologicals and energy (fish and molasses, etc.). Good tilth is needed to facilitate movement through the soil profile. Biologicals were used to till the subsoil and break up the hardpan, encouraging leaching of soluble calcium and sodium. The sodium/calcium ratio also shows the movement through the soil.

* SALT CATIONS - PPM								Ratio	
K		Na		Ca		Mg		Na /	Na /
H_2O	CO_2	H_2O	CO_2	H_2O	CO_2	H_2O	CO_2	Ca	Mg
34	50	90	142	72	396	15	65	2	10
27	71	108	164	73	1744	14	206	2	12
22	59	132	192	61	2240	14	210	3	14
28	61	148	232	72	2136	15	200	3	16
20	48	107	198	86	276	20	46	2	10
13	22	281	352	112	516	26	91	3	14
18	23	267	360	102	816	27	265	4	13
17	29	252	338	87	2000	25	271	4	14

Soluble calcium is part of the salt problem and indicates some drainage occurring at the 4' level but shows a drainage obstruction at the 2' level. High soluble calcium at specific levels shows a hardpan problem.

The degree of movement through the soil is shown by the Na/Ca ratio. 3=slow, 4=very slow and 6=no movement of air, water or roots through the hardpan

High Salt
& Sodium Levels

	Field	Text.	O.M.	CO₃	pH	Salts E.C.	lb per ac NO₃	P₂O₅
1	FRONT 0-6"	5+	0.40	VH	7.3	1.82	20	37
2	" "	5	0.40	VH	7.0	3.36	13	42
3	" "	5+	0.40	VH	7.4	1.73	24	58
4	FRONT 6-18"	5		H+	7.3	1.10	16	37
5	" "	5		VH	7.2	1.44	16	55
6	" "	5+		VH	7.6	1.15	21	36
7	FRONT 18-30"	5		VH	7.4	1.58	25	28
8	" "	5-		VH	7.8	1.44	16	27

The salt level (E.C.) is highest in the topsoil layer, the top 6".

Overview: This is a new lawn around a new house in a new subdivision built over an old citrus grove. New topsoil was brought in for the lawn. The lawn started to dieback after planting. Sampling of the subsoil at 6", 6-18" and 18-30" depths were taken on July 11, September 11 and October 30th. Irrigation was by sprinkler set to apply ¾" to 1" per week with no additional rain.

Recommendations: The soil analysis shows high sodium content and poor internal drainage. Increase the tilth of the soil to leach high sodium from the soil. Biological activity was increased through the addition of soil inoculants of energy (fish products, molasses) and humic acid.

SALT CATIONS - PPM								Ratio	
K		Na		Ca		Mg		Na /	Na /
H_2O	CO_2	H_2O	CO_2	H_2O	CO_2	H_2O	CO_2	Ca	Mg
48	84	184	362	145	1426	20	125	3	18
55	93	105	326	311	817	72	165	1	5
48	90	135	296	105	278	18	143	3	16
29	90	107	412	113	369	13	130	4	32
41	87		328	100	1061	22	148	3	15
27	54	106	235		1375	11	135	3	21
14	39	286	458	101	368	15	163	5	31
12	23	108	218	106	1512	16	160	2	14

The process of leaching sodium (and other) salts through biological tillage can be very rapid with good moisture, air and water penetration through the soil.

The amount of soluble calcium increased once the biologicals went to work. The calcium moved extractable sodium off the soil particles, making it soluble and able to leach away. Noticeable sodium leached through the soil to below the 30" depth in three months or less.

High Sodium, Low Potassium & Calcium Levels

Field		Text.	O.M.	CO₃	pH	Salts E.C.	lb per ac NO₃	P₂O₅
1	W - 1'	1	0.23	0	6.5	0.96	45	87
2	2'	2		0	6.6	0.86	24	18
3	3'	3+		0	6.9	1.01	9	5
4	4'	4-		0	7.4	1.25	9	5
5	C - 1'	2	0.23	0	7.2	1.01	24	48
6	2'	2		0	7.3	0.58	8	27
7	3'	3-		0	7.3	0.67	7	5
8	4'	4-		0	7.7	0.58	7	5
9	E - 1'	3+	0.35	0	7.3	1.15	21	34
10	2'	4+		0	7.6	2.40	8	4
11	3'	5-		VL	7.7	1.78	13	2
12	4'	5		H	7.9	2.02	23	5

Overview: This is a citrus grove without a good biological program. Soil samples were taken of three areas at 1' increments down to 4' in depth.

Recommendations: Sodium levels are high throughout the profile and there is not good leaching through the soil. There is also a lack of calcium, which reduces the amount that is available to exchange with the sodium held on the soil particle. Once the sodium exchanges with the calcium, it is in a water-soluble form and can leach. The deficiency of calcium and the lack of humus reduce the biological activity in the soil and the indepth profile is not improving. Nitrogen and phosphate are being kept at adequate levels through the use of a regular chemical fertilizer program, especially in the top 12" of soil. Below the topsoil levels (at the 3' and 4' depth), nitrogen and phosphate levels drop off dramatically. Potassium levels are consistently low and not up to what are required for a citrus crop, since more potassium is removed than nitrogen during fruit production.

SALT CATIONS - PPM								Ratio	
K		Na		Ca		Mg		Na /	Na /
H_2O	CO_2	H_2O	CO_2	H_2O	CO_2	H_2O	CO_2	Ca	Mg
52	67	164	194	36	67	15	35	5	13
35	54	165	200	31	59	15	32	7	13
25	40	174	202	22	79	11	36	9	18
22	31	182	211	37	88	9	52	6	23
48	58	175	204	40	94	10	46	5	20
36	69	136	189	20	81	13	42	10	15
38	34	156	203	37	63	21	33	6	10
34	51	140	198			32	39	3	6
44	53	198	235	96	02	39	39	2	6
53	64	225	247	102	198	41	88	2	6
29	47	216	238	63	193	27	92	4	9
27	44	213	235	86	503	35	180	3	7

Potassium levels are consistently low and not up to what are required for a citrus crop.

Sodium is high all the way down the soil profile at all sites.

A lack of calcium in the soil reduces the amount of sodium that can leach.

High Humus, Nitrates & Sodium Levels

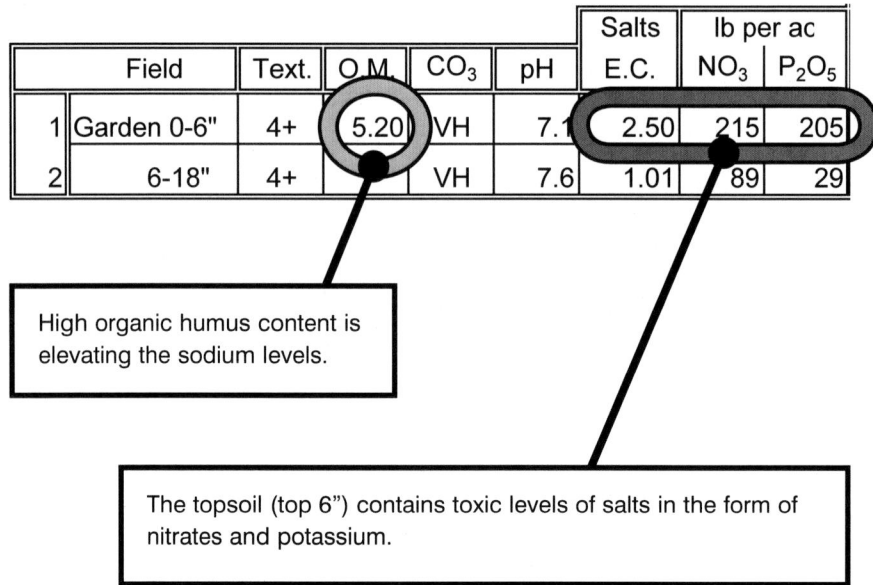

	Field	Text.	O.M.	CO₃	pH	Salts E.C.	lb per ac NO₃	P₂O₅
1	Garden 0-6"	4+	5.20	VH	7.1	2.50	215	205
2	6-18"	4+		VH	7.6	1.01	89	29

High organic humus content is elevating the sodium levels.

The topsoil (top 6") contains toxic levels of salts in the form of nitrates and potassium.

Overview: This is a home vegetable garden that has been following an organic growing program, but without a soil analysis. The plants were doing poorly. A heavy rate of compost was applied with a beneficial high humus (O.M.) content.

Recommendations: The soil analysis shows too much of a good thing. The topsoil shows toxic levels of salts in the form of excess nitrate and potassium. Both harm plant roots and inhibit germination. The compost has resulted in excessive sodium levels, although a good portion is soluble and will leach with flushing. Tests also indicate poor tilth and high Na/Ca/Mg ratios.
Biological tillage is recommended through the adding of soil inoculants and energy. Repeat applications at two to three week intervals may be necessary to improve the soil. Deep tilth will increase deep watering and reduce the soluble Ca and Mg. Subsoil (over a 6" depth) tests show lower levels of nutrients. Deep spading (physical tillage) is recommended to move nutrients downward.

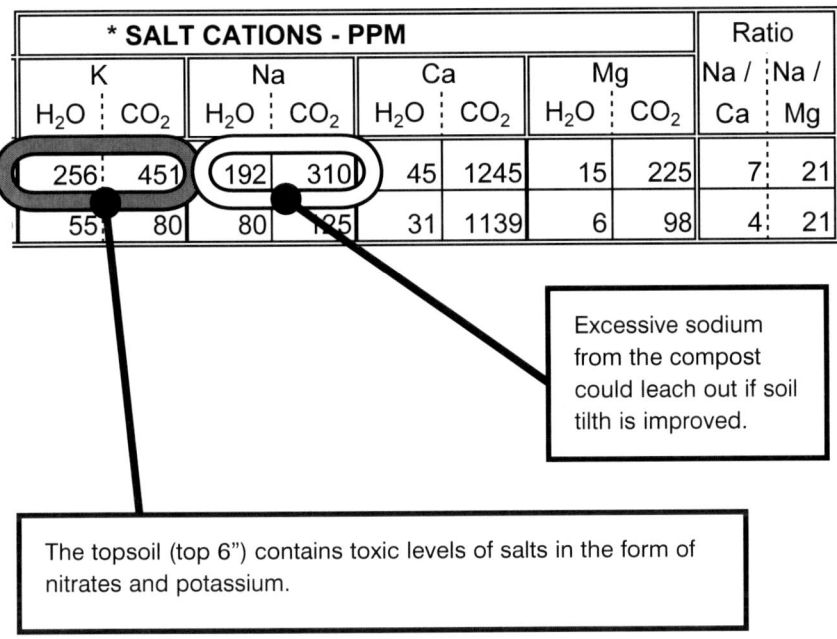

* SALT CATIONS - PPM								Ratio	
K		Na		Ca		Mg		Na /	Na /
H₂O	CO₂	H₂O	CO₂	H₂O	CO₂	H₂O	CO₂	Ca	Mg
256	451	192	310	45	1245	15	225	7	21
55	80	80	125	31	1139	6	98	4	21

Excessive sodium from the compost could leach out if soil tilth is improved.

The topsoil (top 6") contains toxic levels of salts in the form of nitrates and potassium.

Notes on Corn & Other Single-Fruiting Crops

Notes on Corn

Corn, the number two political crop, demands a few paragraphs all its own. The theory that corn and wheat and other storable commodities from 80-82% of the harvested acres are prime movers is based on the parable of the ship at anchor. When the tide comes in, the great ship is lifted by the water, but so are the canoe and rowboat and fishing trawler. Corn is kept cheap by world prices, and this cheapness is translated into cheap feed for confinement feeding, cheapness covering up the defects of that system, which the primary producer can do very little about. However, he or she can trundle home the efficiency in production that "asking the plant" can account for. In one of his educational pieces, Chandler uses a little artwork to illustrate the interrelationships between direction sequences handed out to growers.

Corn-Sorghum Maximum Economic Yield

Soil tests are used to evaluate soil for nutrient-supplying capability and to guide preplant application or corrective actions during the growing season. A representative composite sample of the soil from the root zone is essential for a good soil test. Be sure to take a slice or core of soil 0-12" from several areas in the field, mix thoroughly, and send a half-pint to the lab. Make it representative of the plant root zone area. The subsoil (12-24") should also be tested when maximum economic yield is expected from best management practices.

Plant Testing

There are three critical stages in corn (grain or sorghum) plants.

Less than 12" height — At 25 days, analyze the entire plant for N plus minerals (P, K, Ca, Mg, Na, Zn, Fe, Mn, Cu). Pull 15 plants with roots and wash off all dirt before wilting. Air-dry and place in a paper bag (not plastic) for shipment to the lab. Take plants from a representative area of the field, one plant for every fifth row in a diagonal pattern across the field, starting 50' inside the field and away from the edges. Do not mix problem areas. Sample problem areas separately. At this critical period, before plants form embryo in the ear, essential nutrients can be adjusted to affect yield potential. Apply a balanced formula foliar fortified with deficient nutrients and adjuvants (growth aids, humic acid, microbes, energy (molasses), etc. Help overcome weather stress with plant growth hormones & enzymes. Only minimal rates are needed when sprayed directly on plants. Note: by the 5-6 leaf, embryo in the ear is forming and a balanced nutrition adds rows of grains.

At Boot — This stage occurs when less than 15% of tassels or heads are peaking out. Pull 12 leaves from the most recently mature leaf — when the full dew-lap (sheath) is showing. Cut at the dew-lap and take the entire leaf including the base of the midrib. Mature leaves will have most of the sheath showing. Take samples from the same area and pattern of the field as the first samples. At this point moisture and all nutrients are entering a maximum usage period. Apply deficient nutrients in water or as foliar spray (especially zinc for N use efficiency and proper plant growth). Hormones can be added to stimulate root and grain growth.

At Pollination — When corn silks are turning dark. The major need at this time is nitrate or urea nitrogen for grain filling. Pull 12 ear leaves for sampling. Use the same pattern and area as previous samples. Take full leaves. N, foliar or water applied at this stage has increased yields by 15 to 20 bu. per acre when needed. Only soil nitrate or urea (applied soil and foliar) works this late. Testing evaluates how well this crop has been fed and aids as a tool for adjustment for next year's fertility program.

At Maturity — At this stage, do a crop scene investigation (visual autopsy).

Key Elements for Maximum Economic Corn Yields

During Early Start — Use good vigorous seed. Aid plants with a small amount of high-P pop-up fertilizer in the furrow with a balance of all nutrients and plant growth stimulators (hormones) such as 1-2 oz. of PGR-IV or equivalent (such as enzymes, microbes, energy & humic acid 1-2 pts./ac.).

At Early Growth — Physiologically the number of rows of grain are fixed during the embryo stage before the plant has 7 leaves (12 inches tall). Prior to this at the three to four leaf stage, analyze for nutrients in the whole plants. Appling corrective foliar application of nutrients, hormones, enzymes, humic acid and other plant growth aids can help plants overcome weather and other field stresses to optimize genetic yield potentials.

At Boot Stage — Analyze the nutrients in the youngest mature leaf. Just prior to pollination apply plant growth stimulators (hormones) plus needed foliar nutrients (especially zinc and other growth enhancers), which can stimulate roots and the entire plant functions to boost yields.

After Pollination — Analyze the nutrients in the ear leaf. When silks first turn brown and the grain is filling, supplemental nutrients, especially readily available N (and other aids) in foliar or irrigation application, may greatly increase yields by filling all the grains on the ear.

Any one of these or a combination of all has been shown to increase profits!

Research is bound to recognize this truth as investigations crawl out from under the thumb of huge companies to rediscover the natural universe. As it stands, the average investigator isolates the subject of study in an airtight chamber, reduces variables to a level unattainable in the field, and is always relying on a database quite at variance with the world in which we live. The invention of the randomized and replicated plot served the statistician, but that universe did not comply with the crackle of corn growing at night in an Iowa field. Data confers credentials only on those who satisfy fastidious administrators.

Corn, wheat and the storable commodities are and remain what they have always been: the hinge on which the great agricultural gate swings. And it has to be made of hardened steel, not brittle pewter. That's what

Corn CSI
Crop (crime) Scene Investigation

As the ear corn shuck matures and turns brown, an adequately fertilized plant will have all its leaves still green.

If this isn't the case, look for the following clues to nutrient inadequacy…

Nitrogen deficiency — This will show as browning leaves with an inverted "V" from the tip down. Notice how high up the plant the leaves are burning.

Phosphate deficiency — Showing as purple streaks on the leaves and base of the stalk.

Potash deficiency — Visible as burning down the outside of the leaves.

Magnesium deficiency — This will be noticeable as yellow streaks, but can be confused with zinc deficiency.

Good Nutrition Clues

Ears — How many kernel rows? Most varieties have 18-22 rows at full potential if they are given early, balanced nutrition. Notice the percent of the cob that has large flats, small flats, rounds, and unfilled kernels. A low number will indicate when nutrients were lacking. Note the size of the ear and the stage of the second ear forming. This indicates a lack of sunlight if small. In this case, is the stand too thick?

Roots — How deep are the roots? What are their volume, color and smell? Is there any damage showing?

Stalks — What are the size, girth and height? Do they stand upright? Are they uniform?

At Harvest — What is the bushel weight and grade quality?

Corn Nutrient Requirements

Table 1. Approximate Nutrient Requirements for Corn Based on Stage of Growth for a Yield of 180 Bushels per Acre

Growth Stage	Days After Planting	Nitrogen lb./A	% of Total	P_2O_5 lb./A	% of Total	K_2O lb./A	% of Total
Early	0-25	19	8	4	4	22	9
Rapid Growth	25-50	84	35	27	27	104	44
Silk	50-75	75	31	36	36	72	31
Grain	75-100	48	20	25	25	36	14
Mature	100-125	14	6	8	8	6	2
Totals	Harvest	240		100		240	

Table 2. Approximate Total Pounds per Acre of Secondary and Micronutrients Required for a Yield of 180 Bushels per Acre of Corn

Sulfur	Magnesium	Calcium	Iron	Zinc	Manganese	Boron	Copper
30	50	40	3	0.5	0.5	0.1	0.2

Table 3. Distribution of Nutrients Removed in Corn Grain and Stover

Crop Dry Matter	Dry Matter Distribution	Nitrogen lb./A	% of Total	P_2O_5 lb./A	% of Total	K_2O lb./A	% of Total
Grain 180 bu/A	52%	170	71	70	70	48	20
Stover 8,000 lbs	48%	70	29	30	30	192	80

Table 1 shows the approximate amounts of the major nutrients needed based on the growth stage of the crop. Table 2 shows the approximate amounts of secondary and micronutrients required to produce a corn yield of 180 bushels per acre. Table 3 is the amount of nutrients removed in the grain and returned to the soil in the stover. The data in Tables 1, 2 and 3 were adopted from the Phosphate and Potash Institute and from Crop Nutrient Needs in South and Southwest Texas by Charles Stichler and Mark McFarland, Texas Agricultural Extension Service. Publication B-6053.

the Chandler approach is all about. If wheat is the king of the storable commodities because it has displaced cotton, then corn is crown prince.

There are three major growth periods for corn — early emerging, tasseling and pollination.

When the plant first emerges from the soil, and reaches 8-15 inches, the embryo of the ear is forming, Chandler explains. "If something limits the genetic potential, then the embryo will not confer its maximum genetic potential, which is 20-22 rows of grain on the cob." Many fields have only 12-14 rows of grain, a consequence of faltering fertility. When that embryo forms, notes Chandler, "we have to come in with a foliar application of natural products — hormones from seaweeds, phosphorus, humic acids, things that make that transfer of nutrients in a manner that prevents stunting the genetic potential at the formative stage."

Indicted in stunting may be a lack of zinc, boron, calcium, nitrogen, or too much nitrogen. When nitrogen goes up, it puts down the uptake of all the other nutrients. In short, every other nutrient has to be a star performer when nitrogen paints the field a deep green. The business of going twice over the field with the anhydrous valve open not only annihilates organic matter, it cancels out appropriate uptake of other essential nutrients. The petiole speaks out in thunderous words. "In the conversion of ammonia nitrate-nitrogen into the NH_2 form, the petiole test nails the full message to the bulletin board."

The next stage is the arrival of a tassel. This means an ear to be pollinated. Just prior to that, in the boot stage, there is a requirement for a good flow of balanced nutrients. "So we take the most recent, fully expanded leaf for a sap test — showing what is in the sap and what is available as micronutrients. At that time, we can put those nutrients plus adjuvants in the irrigation water or foliar spray. The key is to balance up whatever is short so when it comes out, we have the maximum number of leaves to make the maximum amount of carbohydrates to fill that kernel out," summarizes Chandler.

When pollination takes place, the kernel starts forming. "At pollination, we take the ear leaf, and anything missing has to be put in the water or foliar," becomes the codicil as Chandler explains the corn growth process. "You can get a 10-30 bushel increase with a foliar application of urea nitrogen plus adjuvants at that critical stage. A wrong step can result in burned leaves."

This much explained, Chandler concludes, "You have to answer the leaf/petiole's declaration with the right product. There's an educational process involved." It takes an independent consultant not tied to any

Wheat Stages
& Nutritional Requirements

Table 1. Approximate Amount of Nutrients Required to Produce 60 Bushels per Acre of Wheat

Growth Stage	Days After Planting	Nitrogen lb./A	% of Total	P_2O_5 lb./A	% of Total	K_2O lb./A	% of Total
Tillering	0-30	25	24	7	15	11	14
Rapid Growth	30-74	46	44	23	50	57	75
Bloom-MUk	15-93	29	32	16	35	32	11
Mature	93-114	0				0	
Totals	Harvest	105		46		76	

Table 2. Approximate Total Pounds per Acre of Secondary and Micronutrients Required for a Yield of 60 Bushels per Acre.

Sulfur	Magnesium	Calcium	Iron	Zinc	Manganese	Boron	Copper
13	14	11	2.4	0.29	0.38	0.08	0.07

Table 3. Distribution of Nutrients Removed in Wheat Seed and Straw

Crop Dry Matter	Dry Matter Distribution	Nitrogen lb./A	% of Total	P_2O_5 lb./A	% of Total	K_2O lb./A	% of Total
Grain 60 bu	79%	75	71	38	83	23	30
Straw 5,000 lb.	21%	30	29	8	17	63	70

Tables 1, 2, and 3 show wheat without requirements and distribution at harvest between grain and straw. Date for Table 2 is from the New Mexico State University. The data in Tables 1, 2 and 3 are from Crop Nutrient Needs in South and Southwest Texas by Charles Stichler and Mark McFarland, Texas Agricultural Extension Service. Publication B-6053.

Plant Nutrient Progress

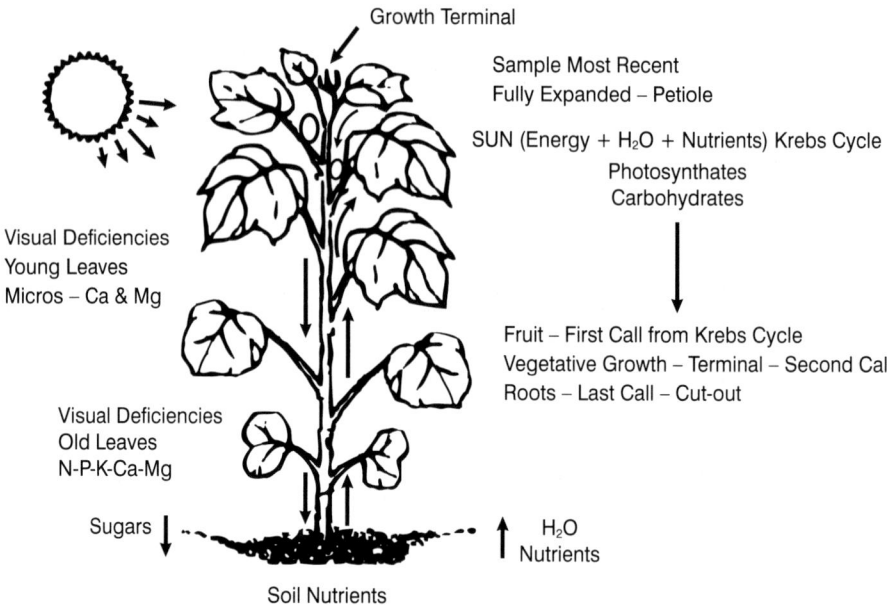

Growth Terminal

Sample Most Recent
Fully Expanded – Petiole

SUN (Energy + H_2O + Nutrients) Krebs Cycle
Photosynthates
Carbohydrates

Visual Deficiencies
Young Leaves
Micros – Ca & Mg

Fruit – First Call from Krebs Cycle
Vegetative Growth – Terminal – Second Call
Roots – Last Call – Cut-out

Visual Deficiencies
Old Leaves
N-P-K-Ca-Mg

Sugars

H_2O
Nutrients

Soil Nutrients

specific product to discover the right fertility aide — this to get the right thing on the right crop at the right time.

Notes on Wheat

Wheat, has four critical stages of growth: tillering, shoot elongation, booting, and early heading. These are very responsive to stimulation with natural/organic products.

The sophistication presented here has to be juxtaposed against the reality that hardly 15% of the farmers even test their soils. Chandler points with discerning honesty to the university people who really deliver, one such being Flake Fisher, Ph.D., Texas A&M. "He was the first one to tell me to put humic acid in our commercial fertilizer, based on Bob Petit's research," Chandler recalls, even though he did not know Petit at the time. "It is the active fraction of humus. It replaces the missing organic matter in most soils."

Fisher told his students, Chandler included, that every time you move phosphorus a half inch from the seed, effectiveness is compromised. If we get our phosphate in the mix, and micronutrients are chelated, the fulvic acid, lignosulfates and other organic materials improve the uptake of those nutrients: molybdenum, zinc, iron, manganese, copper, whatever. Flake worked a lot of that out. Thus, the advice to use starter phosphate nearer the seed, albeit not directly on the seed. At 2" from the root, there is little to no pop-up effect from root zone phosphate.

Wheat is more than the staff of life. It is the king of political crops. Lewis Mumford, the social critic, once wrote that civilization starts with the city. Not so, say the managers of the major breadbaskets of the world. It starts with wheat, and the lessons imparted to mankind via the domestication of wheat were transferred to small grains as well.

Books have been written about the history and economics of wheat. Less is revealed about the art of growing the product. In fact, little has changed during the centuries since wheat left the rich valley of the Tigris and Euphrates and became a world crop. Most of the time, farmers have relied on nature's bountiful soils and regular rain to mix with sunshine for maximum production. As crops began to falter, help came forward. One such help has been Esper K. Chandler. Chandler's message, as codified, lays it on the line with its opening words. "Nitrogen is the major need of all grain plants in the spring." Also, "Nitrogen plus water equals vegetative volume." And, "Balanced nutrition improves quality, grade, and feed value." Chandler tries to reduce to one whiplash line the salient ideas that growers need to know. He never slights the full story even though the short version is more likely to resonate with the grower who needs answers for a political crop. If you ask the plant, it will tell you that it is the flow of energy that governs plant life, as well as economies. It is this flow into the living world that concerns agronomists most. In the beginning, its basic name was wheat, and the instrument is chlorophyll, the green coloring matter in plants. This is the key to natural energy. It transforms solar energy the way a power transformer steps down high voltage for the domestic task at hand. Clearly, the full inventory of energy used to operate a planet, a nation, a people, a wheat field comes from the sun. This is a given, except perhaps in mushroom culture.

The great sociobiologist Edward O. Wilson said there isn't a square inch of the planet that does not support life. The ocean itself may have several times more units of life per cubic inch than any other place on earth, and all rely on solar energy, and so do wheat, corn, milo, and all other crops, political or otherwise.

Wheat Nutrition

Maximize Yields & Grade of Wheat (small grains)

Nitrogen is a major need of all grass plants in the spring. Nitrogen + water = vegetative volume. Balanced nutrition improves quality, grade and feed value. Who puts a steer in a pen with 2 tons of feed and comes back in 3-4 months and ships him to the packer? Yet, conventional wisdom is to feed our crops by applying all of the NPK at planting or soon after.

Phosphorus is the key to fruiting and quality. P in the plant also indicates active root growth at each stage of development. Preplant applications result in only about 10-15% recovery by the first crop. If spoon-feeding is done, recovery is 50-65% — which more than pays for a testing program.

Major Nitrogen Needs start with grain shoot initiation. Apply more N when the first joint is ½ inch (dime size) — maximize more at boot stage. Most plants only require about 10 lbs. of actual N per week, which is possible with irrigation. More than this results in less uptake of balanced nutrients than are required for quality.

Pull 5-10 plants from across each area of similar growth. Rinse off all soil, place in a paper bag for drying and ship to the lab. N and P are the most important for yield. K and micronutrients (especially Mg) are the most important for quality.

Nitrogen efficiency can be greatly improved by the use of soil additives such as natural products (humus, hormones — plant growth regulators, soil inoculants, etc.).

Wheat is especially influenced by soil inoculants, humus products, plant growth regulators, soil inoculants and microbes (these are similar to the special bacteria for legumes). Live, dormant microbes can survive with fertilizers. Some live, active microbes, applied in pure water, can only tolerate some food sources such as humus products and sugars. Check labels for details.

Small amounts of P with N also stimulates root activity for better yields and quality. A few dollars invested properly by using nutrient testing can return three to five times in yield. Ask the plants with a sap (blood) test and listen to them, they use sign language.

Chandler takes the lesson at hand out of the rarefied air of academia and asks a question so mundane it turns profound: "Who puts a steer into a pen with two tons of feed and comes back in three-four months to ship it to the packer? Yet we have been taught to feed our crops by applying all of the NPK at planting or soon thereafter."

Chandler's message to the wheat grower is terse and to the point. "Phosphorus is the key to fruiting and quality. Phosphorus in the plant also indicates active root growth at each stage of development. Preplanting application results in only a 5-15% recovery by the first crop. Spoon-feeding accounts for 50-60% recovery." This efficiency factor more than pays for the cost of testing, and delivers benefits that an extension of these Chandler ideas would duly account for.

Nitrogen needs start with grain-shoot initiation. Here, Chandler does more than walk the soil. "Slit the young shoot," he states, "and apply more nitrogen when the first joint is a half inch (about the diameter of a dime), with maximization of input at the boot stage." Most plants only require about ten pounds of actual nitrogen per week, which is possible with irrigation. To avoid leaving crop performance to chance, Chandler asks his clients to pull five to ten plants per area of similar growth. "Rinse off all soil, place in a paper bag for drying, and ship to the lab." Everything depends on a full plant analysis. As measurements are taken, N and P come out as "most important" for yield, with potassium and micronutrients governing quality.

In his instructions to wheat growers, Chandler stresses the use of natural products, especially humates and biological products often characterized as natural/organic. "Wheat is especially amenable to soil inoculants, humus, and plant growth regulators," he states. "Inoculants should be similar to the multi-microbial products used in soybean inoculation. These bacteria can be live and co-exist with fertilizers." As a parting shot, he notes that "small amounts of phosphates with nitrogen and natural adjuvants stimulate root activity for better yield and quality." It matters not whether the wheat is spring or winter, or any one of hundreds of varieties grown worldwide.

Chandler's protocol for wheat is also the protocol for small grains — cotton, corn, soybeans, and grain sorghum — and most aspects of specialty production if petiole or leaf analysis is performed. Chandler always stresses the learning process that must attend the transition to biologically correct farming. Always, he stresses petiole/leaf analysis and crop-logging on a regular schedule with purpose well defined.

Always a pragmatist, Chandler finds that science merges well with the objective of economical plant production from soil and fertilizer.

"There are no magic formulas or miracle products. All are required to comply with scientific analysis in terms of the exact soil being used, water and crop conditions, and better management practices," Chandler communicates to one and all. "Ours is a long-term learning program that invokes crop management techniques and skills with better returns on investments each year used."

Increasing Yield of Corn & Wheat

Agricultural historians still debate whether wheat or corn rate top historical billing. The land now known as Iraq is recognized as the birthplace of grain production. Ironically, Texas Plant & Soil Lab participated with the USDA and the 1st Cavalry Division in a wheat crop restoration project in Iraq that failed due to bureaucratic inefficiency. Wheat, in fact, was an international crop before there were nations, and the diligent work of plant breeding proceeded long before there were universities to teach the art. The art of feeding the soil to feed plants was well established in Homer's time, but refinements of that art are still continuing to this day.

Chandler's brand of field-applied agronomy offers a learning curve for growers in a world forever on the hunt for a magic bullet. This, when there is a whole range of bullets that science has not even looked into yet. Corn is the one crop that exhibits production potential beyond the dreams of possibility. Twinning, even three ears per stalk, is evidently possible, even practical, albeit elusive.

High-density corn population can deliver more stover than ears per acre, but "how much of our resources should we be willing to spend to achieve the maximum grain and/or silage performance?" asks Chandler. That question about production goals suggests more than one approach to other production programs. How many nutrients and how much water are we tying up in stalks before the factors of production run into that leaf, and the photons of energy process takes place? Energy must be stored in the grain — and more than one ear of grain on that stalk produces maximum economic yield (MEY). The whole picture always remains in question. The main source of energy is sunlight.

It hits that surface of the leaf and ignites all those miracles that go on within that leaf. "If we look at the physiology of that plant," Chandler notes, "and plant growth hormones, activators . . . there are so many compounds, it dazzles the imagination. We have looked at three or four major hormones, but there are several dozen others we have not even looked at yet."

This is not to say that independent investigators are not looking, but often hormone/enzyme-type products are being overlooked. Activators have arrived on-scene. Natural products can stimulate the overall functions of that plant physiology, but it all goes back to the sun's photons of light reaching that leaf. In turn, the leaf has to have the proper moisture and flow of nutrients so that the photosynthate factory can function.

Chandler admits that it puzzles him. "When I look at the Krebs cycle, compounds, sugars, starches, enzymes, as well as hormones — my head swims." A head that swims often calls up memories of events occurring before the arrival of the present impasse. In Chandler's case, it was a reminder that the farmer's job is to work with Nature, not against her. The sheer complexity of Nature's system knocks into a cocked hat the idea that our brand of research science has disposed of all the variables. We are nonetheless impelled to influence the demands and foibles of nature. And so the esoteric formulas of natural/organics beckon, even though the pursuit of economic yield stays on course, moving straight ahead.

The prep sheets that Chandler and his associates put out are not filled with doctrinaire imperatives divorced from bottom-line reality. Chandler cautions, "We can't keep buying inputs for the fields without reference to what the stuff is doing to the bottom line. You see corn being grown in the Higher Plains where water is a finite resource. The Ogallala Aquifer is going down. In 10-15 years we'll be searching for water. At the same time we have the sunlight units up there as well as growing conditions that enable those 300 and 400 bushels. Even so, there is a wasting away of resources with overpopulation of stalks, and overuse of water. In specific areas it may be good arithmetic — but poor economics."

Unlike the government, Chandler never underwrites failures. His forté is called good practices. There are cross-currents. In the case of the High Plains and corn, the saving factor has become drip-irrigation, a system that becomes an inherent partner with the balanced nutrition that Chandler stresses, monitored by leaf and petiole analysis that proceeds almost as routinely as a dairyman taking a somatic cell count.

"Science-led and spoon-fed" might well serve as a Chandler slogan, along with "ask the plant." "Spoon-fed" and "ask the plant" are more than wishful thinking. It takes discipline and a desire to bring the problem into compliance with his own objectives, which, more often than not, break traditional practices.

Chandler remembers the basics and he tags them with the term "efficiency." Mass inputs do not constitute efficiency. Fence-row to fence-row farming does not square with maximum economic yields. Chandler never works out a crop equation without a full appreciation of the fixed costs and the variable expenses. Usually this means increasing yields with available inputs, in effect testing earlier yields, not with "get bigger," but with "get smarter." The real function is to bring the farm back to a natural state, one in which the natural nitrogen cycle works to the maximum possible, and one that serves up a natural carbon cycle as a working mechanism.

Notes on Grains

A common belief on the High Plains is that grain sorghum has iron deficiency. The apparent solution: hit the crop with foliar iron. But, "equally severe or even more severe is the zinc problem. Side by side, it was determined that the perceived iron shortage was actually a zinc deficiency. Zinc limits production more than iron, a sti ll unrecognized conclusion," says Chandler, "and right behind zinc is the manganese deficiency. Concerning watermelons, once the growing ground is prepared, Chandler says, "We can use TerraClean, a hydrogen peroxide non-residue sanitizing product. We inject that into the drip-irrigation system to clean the soil. Then we use a soil inoculant composed of naturally occurring, beneficial, dormant, live microorganisms, a long shelf-life product known to break up hardpan for root penetration and other beneficial actions."

Concerning corn, USDA researchers found that 35% more roots penetrated below two feet after a product containing super microorganisms was applied than with deep plowing, chiseling, or wetting agents. Humic acid is a carrier that mobilizes the nutrient food injected into the drip line.

Notes on Onions

The anatomy of the onion plant is simple in the extreme. Basically, $N + H_2O$ = vegetative volume. Too much N too early equals large necks and leaves, the causes for disease and insect response to the imbalance . . . slow maturity becomes the bottom line. On the other hand, too little too late reduces both yield and quality. Most Mexican and South Texas

Onion Nutrition Sheet

Start onions with high P + humic acid + microbes near to seeding for fast uniform stands.

All plants, like animals, feed every day. Yet many feed crops by applying most of the N and all of the P and K before planting which results in only 10-25% of the P getting utilized. Spoon-feeding P according to a crop-logging program of petiole (leaf) tests can result in a 50-60% recovery of P in crops. At today's fertilizer prices this would alone pay for a testing program (with many more benefits of better yields of higher quality).

For onions, N + H_2O = vegetative volume. Too much too early = large necks and leaves causing more disease and insect pressures and slow maturity. Too little too late = reduced yields and lower quality. "Ask the Plant" with a regular program of leaf analysis and apply as needed. Use the right amount when plants show need to increase yield and quality. Use the right kind of N and apply only on the fields that need it — 10-15 lbs./ac. of actual N/week is the most plants can properly use as a general rule.

Ask the plant when and how much is needed.

Plants feed every day. Why over or under feed? Nutrients can be effectively applied to soil, with water or as foliar feed. Soil fertility and plant nutrition programs recommend minimum inputs to balance needs and get crops off to their best possible start. Then feed in small amounts as needed. Irrigation (drip) allows for the most accurate control on a weekly basis.

Onions Preferences

- Onions respond to only 15% of P at germination — so use more later.
- Humic substances + hormones + microbes enhance P uptake.
- Onions prefer ammonia sources of N.
- N is required at bulbing in small increments.
- Onions respond to sulfur on alkaline soils making sweeter bulbs.

onions are now grown with drip-irrigation only. With the Texas Plant & Soil Lab "Ask the Plant" program of bi-weekly leaf analysis, higher yields of higher quality are produced by spoon-feeding what, when, and how much is needed in a precision ag program.

Chandler's response to these anomalies is to ask the plant. This asking takes place via leaf analysis. The indicated response is application of nutrients only on the field that regularly logged testing identifies. The protocols call for about ten pounds per acre of actual nitrogen per week, this because that amount is the most the plant can efficiently use in that time period, as a general rule. Adequate phosphate each week along with a balance of other minerals is also a must for maximum economic yields.

DNA & GMOs

"Truth is a child of time," Chandler agrees. "It takes time for truth to assert itself." As the harsh technologies visited their destruction on organic matter, tilth, soil physics, and biology, all elaboration seemed superfluous.

It was the modeling of DNA that finally lured scientists and their corporate patrons into that nightmare of bad science. DNA is a blueprint of sorts. It tells cells how to divide and reproduce copies of themselves. Picture a twisted rope ladder. All DNA structures are shaped in this way — those of a dog, a flower, a human being. The rungs of the ladder are made up of four components: adenine (CHN), cytosine (CHNO), guanine (CHON), and thymine (CHNO). These are usually written as A, C, G and T. A can only pair with T, and C with G. Base pairs reproduce themselves, and this is where genetic manipulation enters the scene. Millions of these base pairs form genes. Evolution has taken up the chore of directing the base pair reproduction, frequently and even usually improving the life structure. Genetic engineers have learned how to add and delete from this ladder.

It was a small step to discover naturally occurring enzymes that act like molecular scissors for the purpose of adding or deleting rungs from the DNA ladder. Breaking the molecule has been applauded because of the potential for fighting hereditary disease conditions. Thus was born the idea of cutting and recombining at the molecular level. Thus was born the idea of finding a trait in one organism and transferring it to another organism. Thus also was born the idea of engineering the totality of life.

Two systems are before the world. One seeks to muck around with DNA, to interbreed species of plants and animals at the molecular level, to rescue mistakes with ever —more —powerful chemistry, to come and salvage rather than cause nature to reveal her secrets. Irradiation, not purity, is seen as the key to shelf life.

The second system does more than pay lip service to the conventional topics of humus, organic matter, mineral uptake, tilth, water conservation, line breeding, and humane animal husbandry. It seeks participation in the creation process so that future generations will inherit the land, a land improved, not degraded . . . productive, not degenerated.

The soil scientist watched these developments with alarm as he increasingly embraced the precepts of sustainable farming. It was troublesome to see soybean and corn swept along on the tide of what appeared to be bad science. A few fragments of research called the genetic engineering concepts into question. A Rowett Research Institute scientist found organ damage in test animals fed on genetically modified potatoes. A few farmers reported feeding problems as well as the swine reproduction failures. Growers were promised immunity to pests and disease. It was reported that toxic wastes would soon be degraded courtesy of genetically modified organisms. Natural/organic growers were promised new crops that complied with the standards of the trade, crops that erased the need for pesticides and herbicides, as well as fertilizers. Moreover, as the genetic blueprints of the standard genetic code were unraveled, the scourges of the past would be no more than a bad memory. The government gave its imprimatur to this fiction while Monsanto swiftly took over the seed business and whole chunks of the storable commodity grain trade.

Half the world rejected GMOs, but not the United States. Politicos with no knowledge of the subject endorsed the process.

In a farm world where 80 percent of the soybeans grown are genetically modified (in most American states, at least), simply stated, genetic engineering blends bacteria and viruses for the purpose of creating new combinations. Further lab techniques can then be used to make copies to introduce these genetic materials into organisms, that is, into cells of corn or into embryos of animals in order to make genetically modified cells. In plants, cells are regenerated to start a transgenic line. In the case of cows and sheep, foreign genes are inserted into the embryo or egg to grow a transgenic animal.

The problem with the technique is that it is totally unreliable and uncontrollable. This foreign splinter of DNA ends up in the genome and becomes scrambled. "We don't know at all what they're doing, and they admit that they don't either," say scientists who have banded together for the purpose of seeking a world free of genetic engineering. The scrambling is so bad that scientists can't even sequence the identifying genome. For this reason, the engineered lines are unstable and subject to being re-engineered year after year, a ploy made possible by laws and company clout that requires farmers to "save no seeds," but buy only from the primary supplier.

No one can claim that such crops are superior in nutrition to nature's bounty. Big business argues — backed by considerable advertising — that genetically modified DNA is a carbon copy of natural DNA. Indeed, bad science in big business calls the strange new alchemy "the ultimate molecule."

In Chandler's view, asking the plant complies with the rule of reasonability. Nature's signals are loud and clear, if such a metaphor can be excused. Genes that take off and reproduce themselves in a manner so scrambled that the amino acids can't be sequenced suggest a validity to the term that many natural/organic folk use to label the end product: Frankenfood.

Across the Rio Grande Valley, Chandler can point to growers who outperform transgenic harvests while growing quality crops. This suggests that the spin — supporting GMOs are merely a tempest in a teapot, albeit a lethal one. It presumes to leave soil and plant malnutrition in place while producing bulk, having duplicated the banker's miracle of creating money out of thin air, in the case of transgenic crops engineering plants to withstand starvation.

This observation leaves unexplained how the merger of bacterial and viral aggressors in the GMO process promises new combinations quite capable of building new viruses and bacteria that cause diseases resistant to all known treatments. "Bad science . . ." intones Chandler. "There's no place for it in quality agriculture. That's been proved with white wheat and yellow rice. The bushels may be greater, but the feeding quality subtracts the gain in a big way."

Notes on Watermelons & Other Continuous Fruiting Crops

The Lower Rio Grande Valley has long been recognized as a specialty crop heaven. Snowbirds that winter in South Texas, fondly called Winter Texans by locals, see the fields of citrus, sugarcane, corn, milo, cotton, soybeans, watermelons, cantaloupes, strawberries, peppers, artichokes, onions, celery, lettuce and various greens — indeed, a count that seems ever to increase, which is precisely the point of exhaustion for heavy front-end fertilization.

Notes on Melons

Watermelons signal a calcium deficiency by cupping their leaves, this in contradiction to lab readouts. Chandler looks pensive. The lab does not lie, but it misleads when a researcher relies on single-factor analysis. In this case, "we have low boron. Boron is extremely important for calcium metabolism and carbohydrate transport. You can uptake calcium and yet not metabolize it for want of boron." Boron nutrition suffers from the stigmatisms of outdated concerns of toxicity with overuse.

Under Texas circumstances, it is worthwhile to treat the watermelon field with TerraClean to kill soil pathogens and promptly repopulate with beneficials. The chemical equation is slowly being replaced by a bio-norm. Compost and humates, once a mainstay of the natural/organic underworld, are now front-line stuff in the vegetable and watermelon fields. Watermelon grower Nowell Borders has his trucks hauling in raw materials for non-traditional humus products. Thomas Harr uses feedlot manure, field refuse, corncobs, fish waste, poultry litter, and other manures to make compost. Some food safety bureaucrats in their ignorance of soil microbiology are targeting manure as a hazard in food-producing crops.

Green sand out of East Texas has also made the transition from backyard gardening to commercial watermelon production — ditto for Phil Callahan's paramagnetic rock. Commercial caches of rock powder from Nevada are now moving into the fruit and potato industries. Humic acids and fulvic acids have also been established in a farming industry that still eschews organics.

Nowell Borders' acres look to innovation, and innovation often harks back to small growers who weren't satisfied with companion planting, and expanded their answers into solid findings. "I want to make this soil better," says Nowell Borders, who is probably the biggest watermelon grower in North America. He has the innate ability to see and develop total programs of production and marketing with innovative products and methods and they for the long-term quick fix. His modestly large operation exhibits the classic Chandler approach, all of it refined by client-consultant conferences as the season proceeds. Nowell, with Gene Satterlee, pioneered watermelons on drip in South Texas, allowing their growth on clay soils.

Thomas Kyle and Bubba King work for Nowell Borders, whose watermelons are grown under sustainable rules for premium markets, and under standard protocols for commercial markets. Use is being made of natural/organic products and rotations that so far have created a model system, and Tom and Bubba, under Nowell's direction, apply them all, to the best of their knowledge. The mood in the countryside is to tiptoe into useful procedures. Indeed, many a farmer who has "gone sustainable/natural/organic" can't tell you when it happened, the transformation having been that gradual. Sudden death transitions often mean just that: sudden death.

The former Starrco Produce of Rio Grande City was the first large-scale Texas pioneer of drip-irrigation. They used the lab and Danny Sosbee, a fertilizer supplier and composter, to formulate drip-compatible plant foods.

"We have used several soil inoculants along with a couple of live organisms. It has a lot of anti-pathogens in it," Kyle says. Often the natural/organic products are locally produced, and hardly known outside the region. The one Kyle describes, F-68, has been made in Mercedes since 1968 with hardly a university being aware of it. As with so many natural/organic products, this one emerged from a family background of herbal healing known locally as Curanderismo. It takes an open mind to look past double-blind studies by credentialed folks to "what works," as the watermelon grower puts it.

Cultivation of the Borders' acres calls for every other row to serve as a windbreak, usually a small grain — spelt or triticale being the most popular. Sunflowers may be a better companion plant except too much sunlight blockage would be deleterious to watermelon growth. Other windbreak crops of corn, giant mustard, etc., are being tested. The Rio Grande Valley has tremendous winds. These shift the watermelon vine and shed the oncoming fruit. The small grains are turned under along with the vines when the crop season is finished. If the small grains begin to compete with the watermelon roots, they are undercut and annihilated. Some growers use hormone sprays, a practice fraught with danger. A drift to the watermelon patch can be lethal. A short-stem rye grain that dies out early won't compete with the watermelons. Triticale keeps growing as long as it can partake of drip-line water availability.

Top-yielding watermelon growers use two drip lines, one in the middle of the windbreak, one under the row. Nowell Borders was one of the first to invest in an extra drip line. Seedless watermelons are interplanted about every fourth plant or according to some other variation: one in three or five, etc. Sprig 'em or seed 'em! But transplanting allows for a head start.

The watermelon crop is a poster child in terms of fleshing out the Chandler "ask the plant" procedure. Start with the 12-inch samples in 12-inch increments to inventory plant nutrient reserves. Noel Garcia, one of Chandler's good right hands, explains, "Precision sampling is needed, but not over 20-acre size samples because soils can vary widely." Those who see what they look at are urged to note color, shape, slope, texture, and plant growth. Soils always test differently in terms of averages, not samples. When tests of the watermelon ground so suggest, various areas will require different treatment. Special fertility rates are indicated for spot remedial treatment areas, and then an average program for all. "Small precision samples have to be averaged for ease of management," Chandler reminds. "This is for managing whole fields or irrigation zones."

Plants ask for feeding according to physiological needs. These vary greatly during emergence, the vegetation stage, fruit forming, and the maturity stage. Field information forms sent to the lab, with samples, are necessary as the growing chore gets under way.

In commercial production, Chandler counsels that start-up should be with little nitrogen and phosphate. It is counterproductive to feed too much too soon. Adding more of these nutrients as the plants exhibit need defines both the procedures to follow as well as the outcome. One must be absolutely certain that there is adequate humus, calcium, magnesium and potassium throughout the soil. These can be broadcast before planting, side-dressed, added in water, or presented as a foliar. Micronutrients such as zinc, manganese, boron, copper, iron and molybdenum can be tested for deficiencies and applied foliar, if needed, or to the soil. Always ask the plant is the advice of Chandler, the lab and the plant itself. Crop-logging is mandatory, an absolute that walks with almost all production, specialty or otherwise.

How do you ask the plant how it feels while it is growing? Here, a review of paragraphs on petiole analysis is in order, albeit with the peculiarities of the watermelon in mind. Indeed, the petiole test defines the needs of the plant according to the genetic potential of the plant. The term "specific" comes to mind when petiole sampling is utilized. Nutrient contents of petioles and leaves vary with location of the leaf on the stalk. The age of the leaf also makes a difference.

Petiole nutrients tested from the sap of plants tell of future growth that can be expected to arrive in 7 to 21 days. Nutrients that have been utilized by the plant to date call for a postmortem leaf test. For this reason, and for reasons that become self-evident as the watermelon field inches its way toward fruition, there is a requirement that an exact inventory of information as to sap or leaf supply and samples be supplied. Chandler's tests differ from the old, for which reason proper standards can guide to better fertilization. "One leaf per plant," is the terse directive, "and the exact same age and location for each of the composite samples. Take 30 to 50 individual petioles. Numbers depend on the size of the leaf or stem."

A bigger sample results in too much volume in the lab and can result in sample segregation. Too small a sample results in poor aliquots. The lab is quite firm about total compliance with proper handling. The moisture level must be dried slowly and cleanly. "Find the main runner, then count back from the most recent fully unfurled leaf to about the sixth, the most recent fully developed leaf regardless of the count. Do not go back towards the older leaves or upwards toward the younger leaves. For a petiole sample, discard the leaf and send the whole stem to the lab, including a few typical leaves for observation." Field directions are that specific. They continue, "Wash samples thoroughly before they wilt to remove dust and other contaminants such as sweat. At least rinse with clean drinking water before bagging. If recently sprayed with anything, foliars included, use a non-phosphate detergent such as Ivory or Joy dishwashing liquid. Thoroughly wash the surface of the leaf, but only rinse petioles, and do not crush. Rinse at least twice in clean water. The best rinse should be with distilled water. Handle with clean hands and place only on clean surfaces. Use only paper bags. A simple rinse is better than nothing. Leaves and petioles dry best in paper bags. If plastic has to be used, then punch holes in the bag to permit air to enter, otherwise molding in transit can be expected."

Drying calls for a bit of ingenuity. The dash board of a car works with sun heat. Hair dryers work, as does a warm breeze. Fully dried samples can be stored for quite some time without deterioration. Whatever the method, it is best to strive for temperatures not over 160 F for eight hours or longer. Only low heat can serve this slow-drying objective. Directions for the watermelon grower seem like orders that are military tough. Fertilizer history, age and size of plants, fruiting stage, moisture level, insect pressures, plant disease damage: all go into the mix called recommendations. For this reason, a veritable police blotter on the plant has to accompany the plant sample to answer the questions the grower asks, via the petiole. Data on planting, variety identification, et cetera all figure in because both the devil and salvation are in the details.

The watermelon is a very deep-rooting plant that prefers light-textured deep soil. It wants living room for taproots that are forever on the hunt for moisture and nutrients. The plant has a survival instinct in dry weather, and rewards the thoughtful farmer who slakes its thirst in a timely manner. There was a time when all or most of the fertilizer was applied at pre-planting. This drastically limited the plant's potential. A watermelon total of 25,000 pounds per acre was considered very satisfactory if not superb. This outlook led to excess nitrogen and excess vines, diseases, insects, and delayed maturity, as phosphate and mineral uptake faltered. Nematodes and disease seemed to follow the rotation within five years, with soil healing required, not so much because of the plant's drain, but because of excess nitrogen and unavailable, confused, and complexed phosphate. Layout years required as much management as the watermelon crop-producing year.

As irrigation progressed and water conservation was answered with the drip procedure, the door to really superb production was kicked open. Spoon-feeding entered the watermelon producer's vocabulary. Heavier soil became usable. Fewer rotations suggested themselves. Active roots became limited to the zone of moisture provided by drip-irrigation, and the nutrition zone followed suit. That's why yields of 40,000 pounds per acre became attainable. Many growers are exceeding this.

Near College Station, Texas, the Wiggins family, consistently break watermelon yield records. They grow melons in three different areas of the state, and are known in the trade for their top-quality crops. Bob Dyer of Growers Select Produce in Mission, Texas, also a three area quality grower, is gaining recognition as a leading exponent of compost tea not long after adding it to his repertoire. Young growers such as Mike Helle, Wayne Reavis, Benton Beckwith, Glenn Marecek, Ronnie Livingston, Ronnie Skloss, Danny Arnold, Jack Wallace and others adopted sustainable natural/organic inputs after listening to their plants.

For Chandler to consult for the biggest watermelon producer in the nation is itself a mark of distinction. For the technology of Texas Plant & Soil Lab to have a role in creating a system that nature itself could approve of is equally worthy of a prize.

"Regular petiole testing reduces variables to a near-vanishing point." Being able to read the sap pulled from the roots to the leaf is at least as profound as lesser developments now enshrined in the annals of science. Indeed, this ability to ask the plant has delivered 70,000-90,000 pounds of watermelons per acre, and this has been attained with 30,000 pounds or more on the first cut.

Moreover, Chandler has been a leader in bringing non-academically approved products such as humus, humates, hormones, enzymes, carboxyls, lignosulfates, activators, soil inoculants, and naturally occurring paramagnetic rock, plus many types of naturally occurring beneficial soil microorganisms into the commercial growing arena. His primary question is always the same: Would Nature approve? Sometimes Nature does not approve, but growers have absorbed more indoctrination than they can easily overturn.

Legume inoculation is improved when a very small amount of humic acid + phosphorus + energy via molasses + soil inoculants all are added to the legume inoculants sprayed on the seed in the soil. A little is great; more can be disastrous.

The watermelon plant likes deep sandy soil. It wants room to go deep to find water while the melon is developing. Today, however, watermelons are grown on all kinds of soil, good counsel and the grower's art rectifying the shortfalls that nature has bestowed. To some extent, these machinations impose an artificial environment. Even heavy clays are being pressed into service, a measure unheard of a few years ago. In the last case, rows are mounded and covered over with plastic both to assist in weed control and drainage, and to facilitate warming of the soil and preserve water.

It is always the objective of the specialty grower to produce his crop a bit out of season, this to hit the market before growers on colder soils are ready. Almost all the literature held over from the reign of NPK leaves unreported the role of calcium. Actually the watermelon is more dependent on calcium than on NPK. Any time a nutrient stress develops, calcium becomes less available.

Chandler has developed a "CSI" program for crops, modeled after the popular television series where visual evidence and samples for lab analysis are collected. Chandler's acronym stands for Crop Scene Investigation, although the misuse of expensive plant foods and chemicals could also be characterized as a crime against nature. Visual evidence of nutrient deficiencies are noted along with their effects on physiological development of plants and fruits, and a lab analysis of samples dictates sufficient nutrients for the crop. Chandler's CSI also involves the use of the flashlight at night to observe young emerging seedlings. He asserts that the beneficial effects of nontraditional natural product on the plants and roots can be better seen, smelled, and measured in the dim light.

The statement regarding checking the crop invites so many variables that it courts deletion from any production manual. Take watermelons. "You check watermelons weekly," Chandler advises. "It is a rapidly growing crop. There was a time when growers considered 10,000-15,000 pounds per acre a good crop. With fertility management, we got up to 20,000-25,000 pounds, which is where the average grower is today, but that is still not a very profitable crop because the watermelon can produce 40,000 to 80,000 pounds per acre until the weather becomes inclement," explains Chandler.

The art now allows growers to harvest the watermelon crop for 12 to 15 weeks. Formerly, four weeks would see the end of the crop. It is the nutritional balance that presides. Drip-irrigation permits spoon-feeding, a procedure hardly envisioned during the days when factory-acidulated salt fertilizers conquered everything in sight. Balance cancels out disease and insects and most of the weed competition. Profits come from spreading fixed costs over more total units of production. As long as inputs return their costs, the unit cost of production goes down.

Nowadays, the norm for harvesting melons is less than 60 days, but this often stretches to 120 days. Hot winds often intervene. The real challenge is to maintain the equilibrium. Shorter days and lower temperatures close down even the best of performing crops. Light survives through the winter, though in diminished amount. A light and temperature shortfall short-circuits the continuously fruiting vine. If properly fed, it is possible to put a melon on every fifth to seventh node. Hormones, enzymes, and other natural products in the fertilizing mix enable production of carbohydrates. The innovator has baffled the academician. Sugar, applied foliar and/or in the drip water has been known to put a fruit on almost every node. The vine can't hold or control such an overload. The carbohydrate triggers a mechanism that says, "store me," and creates a fruiting site.

Drip-irrigation with all best management practices (BMPs) can keep the root going until the crop maximizes a multi-fruiting crop. Using the tests Chandler relies on, he and his growers have harvested watermelons for 120 days, this when 4-6 weeks has been a normal period for watermelons. The increased return on investment can be computed even without a calculator.

Notes on Potatoes

Each entry in the pantheon of specialty crops asks for special notes aside from the universal lessons that take up most of this book. Potatoes are unique. Every state in the union except for Alaska grows a significant amount of potato crop.

Potatoes create a quandary not easily solved as long as conventional agriculture prevails. The current practice is to slug the crop with nitrogen early, expecting the crop to grow out of the overload. This runs counter to plant physiology. Running up the nitrogen load reduces the uptake of minerals. Then the plants must struggle to get the potash, phosphate, calcium, magnesium, and micronutrient uptake they need, resulting in vegetative volume, not tuber production. Indeed, one of the problems with salt fertilizers is that they become available to the plant outside of the biological system.

Idaho and other states have worked on petiole analysis, but the spoon-feeding of nutrients mandated by such nitrate- and phosphate-only monitoring was considered too irksome. Put out the fertilizer ahead of planting and wait for the tuber crop to consume the nutrients as the season proceeds — that seems to be the conventional game. Sustainable farmers can easily understand how Chandler's approach is more efficient, how it wards off insects and other crop destroyers. Researchers in Idaho and other states are now turning to biological soil inoculants to control soil-borne potato diseases since chemicals have failed to do the job.

Nitrogen is little demanded before the plant starts making the vegetative volume of the tubers. It comes into its own as growth proceeds, adding to the mysteries of plant growth that have still to be defined. Excess nitrogen can aggravate many soil-borne disease organisms.

If something is edible, it's a commercial possibility and has special nuances that attend growth and field or garden management. If the plant rates mentioning, it is probably grown in the Rio Grande Valley. Some specialties are major crops, such as watermelons with celery being also grown in a well-manicured and drip-irrigated field. Much the same can be said of onions, beans, cucumber, icicle radishes and others.

More Vegetable Crops

Turnips, rutabagas and kohlrabi go together with mustard the way salt and pepper pair off with meat and potatoes on the dinner table. Such plants are standoffish, usually hindered rather then helped by botanical interlopers. Not so with potatoes, Irish or otherwise. The potato tuber is at home as a monoculture and as a denizen of a diverse garden or farm setting. So is a mix of beans — bush, lima, even pole — because their root excretions help control insects and fungal crop destroyers. Chandler's notes on feeding the plant as needed find particular application here because commercial potato producers rarely endure the presence of a foreign crop whether symbiotically compatible or not. Broccoli, cauliflower and cabbage ask for the same soil type, the same careful "as needed" feeding, as is the case with corn and fava beans.

Asparagus, chard, carrots, cauliflower, celery, chives, cucumbers, eggplant, garlic, kohlrabi, greens, melons of every stripe, okra, parsley, peas, peppers, radishes, rhubarb, spinach, squash and strawberries — all are commercially viable when the price is right. The rules of production outlined by Chandler preside just the same. There are over 65 commercial crops that grow in the Valley. That is why no one who works for Chandler belittles books like *Roses Love Garlic* or *Carrots Love Tomatoes,* both by Louise Riotte. Here, the most wizened row crop farmer can learn that asparagus finds botanical romance when beets, chard, carrots, lettuce, onions, parsley, spinach, strawberries, tomatoes, along with basil and several flowers mix their auxins with root exudates. No one really knows why. Folklore gardeners just know it as a fact. One day, investigators may put a cause and effect handle on an unseen phenomenon that reaches back almost as far as acupuncture.

Now, let's extrapolate a bit. There are lessons that are specific to particular plants, but they are largely unknown or given reduced status. The gardener will tell you that celeriac and celery, corn, cucumbers, grapes, leeks, lettuce, mustard, melons, peas, pumpkins, radishes, even soybeans, share factors that open pathways for balance and growth. It would take an encyclopedia to explore the nuances that help or hinder each specialty, most of them beyond the purview of a laboratory and outside the agronomic practices of the hour.

Notes on Bananas

Plantation managers and gardeners have always led lives that aren't exactly parallel. The old colonials who grew bananas south of the border followed excellent management practices, but they didn't know beans about nutrition. That statement could bear Chandler's quotation marks without much revision. "Crops that rape the land," says Chandler, "respond to nutritional intervention." In the case of bananas, it is calcium and potash. Management, cultural practices, and harvesting cannot substitute for a lack of fertility.

There are agronomists in the countryside who have mastered visual inspection. "When environmental factors reduce the natural production of carbohydrates, it follows that the nitrate in the plant or sap puts cause ahead of effect," says Chandler. "Soil moisture conditions, sunlight intensity, temperature, the amount of wind, the fruit stage, and the load of fruit — the growth, the vigor, and the shading of the lower leaves — all affect carbohydrate production. The grower's response has to consider how much to put on of the carbohydrate factor. That's where we found that sugarcane molasses, used as a plant and soil adjuvant for centuries, contained the widest diversity of available plant foods. You can use beet molasses, high-fructose corn syrup, sorghum molasses, or you can use table sugar sucrose, dextrose or Sorbitol. You can use them as beneficials. With sugarcane sugar you have a wide range of other nutrients. When visual inspection turns up symptoms, carbohydrate manipulation may be a way to handle it. That's taking response past plant nutrition." Circumstances invite specific responses. Smoke from agricultural fires in Mexico sometimes rob Rio Grande Valley plants of their sunlight. Low light intensity in the midst of watermelon production inhibits leaf carbohydrate production. Foliar treatment with sugar then becomes a salvage operation. Volcano eruptions can deliver the same shortfall.

Techniques invoked by Chandler know no geography. They work as well in the tropics as in Canada, Africa, Europe or Asia. They adjust to the situation. They embrace crops as different as coffee and canola, tropical fruits and Alaskan pastures.

Notes on Coffee

The key to the above statement requires explanation. Take coffee, which is not an American crop, but is a crop often referred to by American consultants. The industry has a manual, a trade bible, which states blandly that coffee does not respond to phosphorus. This would be an astonishing statement if true, which it isn't. Under most conditions, the statement may be true because volcanic soils are very acid. Phosphorus in acid soil ties up tightly with aluminum, possibly the most ubiquitous mineral on earth. Add iron to that acid equation.

If you raise the pH to a reasonable level, say six, the phosphorus response will be excellent. Availability is the signal word, and it is governed by high calcium lime. A lack of phosphorus simply stunts plants, whatever the species, and compromises root development. Without suitable phosphorus, the crop is produced outside the desirable range into a forced environment. Rainy seasons, dry seasons also figure in. Soil leaching runs rampant in the first case, along with micronutrient deficiencies — iron, zinc, copper, boron, manganese.

Dicots, monocots, all have their fingerprints, or leafprints, if you will. A trained consultant can read them from a slow-speed pickup. Even knowing the crop isn't essential, and the Latinate name is simply superfluous. Often weeds at the edge of a field telegraph deficiency symptoms, as explained in *Weeds: Control Without Poisons*. If deficiency signs show up in the denizens of the waste areas and ditch, likely they'll show up in the crop, specialty or otherwise. If Johnson grass has a magnesium deficiency, it can be assumed that the same deficiency will prevail in the crop field.

Notes on Sugarcane

Sugarcane is the granddaddy of all political crops. The second act of the First Congress assembled under the new constitution in 1789 was a sugar tariff, this so a sugar industry could be developed in the new United States. The wage level was 50 cents a day. This made a two-cent tariff a very high levy for the foreigner to overcome. Sugar continues to have a measure of protection into the 21st century.

It is one of the ironies of agriculture that so much is known about consumer sugar, yet so little about how to use it as a crop aid. Sugar is after all a highly available form of carbon. "But first of all," became Esper K. Chandler's wry comment, "we have to produce it." Historically, the Rio Grande Valley goes back to the beginning of the 20th century as a sugar producer, this because of its subtropical climate. This area seemed to fade out of the picture over time. In the 1970s, Texas A&M, with the help of its researchers, studied sugarcane and found that the crop had potential in the Valley. In the mid-1970s, a successful modern mill was constructed at Santa Rosa, a bit north of Weslaco, to accommodate harvesting and processing. Its state-of-the-art technology made it among the most efficient in the world.

Sugarcane is perhaps the toughest water-requirement crop, a fellow traveler with rice. Normally, the crop is grown in high rainfall areas. A come-lately sugar crop, the sugar beet, is generally produced in more temperate climates.

Sugars have classes: sucrose, lactose, dextrose, fructose, glucose. More recently, corn sugars have claimed the grocery industry as their own, and American bodies are bloating as a result. Now, with protection gone and the sugar subsidy for developing nations in limbo, sugar borders upon being an academic subject. Growers do not believe this and therefore the crop stays on, like farmers who consume their capital because they have an underpriced product.

Louisiana has the oldest sugarcane industry. The harvest target is December. Texas literally closed down sugar until the industry was revived in the 1970s. Florida now grows more sugarcane than any other state, but Louisiana, the second largest producer, still claims the title Sugar Capital. Texas has its own Sugar Land even though it produces less than Hawaii.

Most sugar plantations require a laboratory and an agronomy program to deal with fertility and water management. Plant analysis evaluates the need for moisture also. That factor rates front-burner attention along with the nutritional requirements. Chandler recalls,

"Desmond Allies, a displaced English Rhodesian and farm manager of the Groves Enterprises, once plopped down a copy of H.F. Clement's 1980 book, *Sugarcane Crop Logging and Crop Control: Principles and Practices*. It remains possibly the most physiologically correct plant analysis and soil fertility primer available. 'This is what we want the lab to do,' he said. By applying these proven principles, the lab was able to increase yields and the sugar content of their clients' crops well above the local university norm."

The sugarcane plant is unique because water evaluations can be made with a sheath moisture index. This means the moisture in the cane sheath is a constant factor. It hardly fluctuates during the day regardless of heat units. The folds of the leaf act as a reservoir. A readout of sheath moisture makes it possible to schedule irrigation. A caveat must be inserted here. Drip-irrigation has been tried on sugarcane in the Rio Grande Valley, but it was unsuccessful because it was not coordinated with plant nutrition.

"I don't think we've put the whole picture together," Chandler admits. "One fallacy is that water will cure all problems, but water has to be tied to soil fertility and balanced nutrition. This takes us back to tilth, the humus factor, and the water-holding capacity of soil." Humus, of course, improves the cation exchange capacity and feeds the necessary soil biology balance. All these factors, Chandler asserts, dispel the myth of "pour on water and all problems vanish!" The drip has proved itself on many crops, but not without full attention to the variables that forever bedevil the heads-up grower. The lessons expressed here have now been taken up by the drip-irrigation companies. They now tell growers that getting the most out of the water depends on loading that water with the nutrients that the plant says it wants, if only you ask it. The cut-back from pivot and flood irrigation speaks volumes about the role of the lab and about traits, both unidentified and known.

Cane is different. It is a constantly growing crop. Most other crops have a limited growing time. In Hawaii, sugarcane is a two-year crop. In the Rio Grande Valley, it is a one-year crop. In the last case, the objective is to grow the crop efficiently on a one-year basis. The function of a vegetative volume crop like cane is dependent on nitrogen and water. Volume is one thing, quality another. Now minerals and micronutrients come into play, starting with phosphorus for root growth, then potassium because the sugar crop uses more potassium than nitrogen. That also happens to be true of all sugar crops: melons, beets, citrus, etc.

Here the error of strong acid extraction asserts itself. Using that method, the laboratory is likely to conclude that the potassium fix is adequate or more than adequate, but the plant with its weak acid-extraction capability cannot get that potassium fast enough to optimize the sugar potential. Once the potassium level is appropriate, then it has to be balanced with magnesium for the chlorophyll molecule. Next, the available calcium, really the VIP of the cell wall, governs the health of the plant. Added to the above is the transport element: boron. Without boron, transport of carbohydrates from the production site on the leaf surface to the storage facility in the plant simply doesn't occur. If there are carbohydrates left over from the demands of the plant roots, the soil stakes its claim on behalf of the microorganisms. If carbohydrate production is less than adequate because the leaf is not functioning, a short circuit occurs. Hormones, transports, activators, all depend on minute amounts of manganese, copper and molybdenum. If the plant is ill-fed, little remains for the root and the soil's denizens.

The leaf test identifies exactly when the root food supply is giving out. When the root slows down, it stops picking up phosphorus. Academia seems to reject this Chandleresque finding, and goes blithely about its business linking strong-acid extraction and the presumed availability of everything with a cation exchange number. This carbohydrate development function is critical in the sugarcane crop. Measuring traces takes on an urgency because of a growing realization that ambient air furnishes some of those nutrients, the leaf stomata being the port of entry. After the major and secondary nutrients, attention should be directed to studying and evaluating the role of micronutrients and trace elements, much as might be the case with genetic sequence. A few years ago, it was speculated that unless discovery was speeded up, we would have to wait 400 years before we found what we needed to know about the traces. The Genome Project shortened that time frame dramatically, and the Standard Genetic Chart may have identified the missing information which now needs to be plugged into the lexicon of plant knowledge.

"That's where paramagnetism comes in," Chandler says. The supply of lava rock seems endless, as are the minerals in the sea. The best were a part of the ferment that old Jim Martin created as a precursor to Medina.

"Our slowness to plumb this treasure trove reflects man's inability to keep up with nature," Chandler says. His database on paramagnetic rocks represents a collection of samples from across Texas and surrounding states as well as from international sources. Well in advance of university studies, practical growers are testing applications and taking note of practical results. In time, the bookkeepers will no doubt be able to refine a database to prove what the pioneers knew all along. How new findings will impact on sugar crop production will be revealed in the fullness of time. As a data bank takes form, and becomes a tool for consultation, this far-out information will find service in areas now considered non-essential.

Plant-produced sugars — polypeptides, polysaccharides, etc — furnish the microbial world a cafeteria never listed on any fertilizer bag. As a preferred source of food for microorganisms, the unrefined sugar byproducts were returned to the soil by folklore farmers, always with excellent results. The carbon content of molasses is well known. Carbon is also a safer adjuvant to prevent leaf burn from foliar feeding. Feeding unpaid billions of microbial workers is also a lesson that natural/organiculture has given to growers, a lesson some accept reluctantly. Nutritionists are now touting cane molasses for its additional mineral and micronutrient content, and especially for its digestible calcium. Plant growers have long observed these benefits as well as the multi-sugar content that encourages fruit formation on multi-fruiting crops.

The traditional practice in cane production is to burn off the leaves to facilitate harvest. Working with his associate, Bubba King, Chandler dealt with the micronutrient protein in sugarcane fields by using humic acids in an effort to make life tolerable for the microorganisms responsible for extracting many micronutrients that are locked up and unavailable. Cane stover is heavy in silicates that are difficult to decay. Usually a mass three and four inches atop the soil, it impedes cultivation and harvest. Earlier, a farmer would make trip after trip trying to remove the biomass. Nowadays, Chandler's clients leave the mass in place as a mulch and inoculate it with humic acid, molasses, and microbes — conservation tillage, if you will!

Flood irrigation is not impeded by the heavy mulch. In fact, the water works its way underneath the mulch, and then when the mulch settles down and the soil inoculates go to work, the benefits achieve exponential status. By harvest time, the mulch is gone. "You can turn a few heads," observes Chandler, "but the growers who run with the ball are few." Just the same, AgriEnergy of Princeton, Illinois and the local USDA office have picked up on the concept. Each new concept results in new procedures.

Sugarcane, like most plants, needs nutrition every day. If you over-feed the plant or starve it, the crop is the worse for it. But if the leaf is monitored regularly and nutrients are supplied as needed, the resulting crop is light years ahead of a conventional crop. Using natural/organic materials to activate the chemical fertilizers is an absolute necessity. When combined with the "Ask the Plant" program, drip-irrigation greatly improves the economic utility of water and plant food, resulting in higher tonnage and sugar content.

Lessons from Cotton

"Cotton is king." That is, it *was* king before it was reduced to share-cropping and land destruction. Arbitrarily, one has to select the moment in time from which to move forward or back. We might as well go back to 1855 when a man named David Christy published a tome entitled *Cotton is King*. His book argued that the Union had to make peace with the South. Stump orators, Congressmen, and economists all picked up on the slogan. During the debates over Kansas, South Carolina Senator James Henry Hammond thundered, "The North dare not drive the South to secession. Without firing a shot, without drawing a sword, should they make war on us, we could bring the whole world to our feet."

It was one of history's gravest miscalculations. It rivaled the blun-der that saw cotton acres mined to death, then suffer from diminished return until the post-World War II era, when almost everyone reached for a magic bullet.

From soil to laboratory, from planting to harvest, plants are moni-tored the way a physician monitors a patient after a serious procedure. As a consequence, cotton fruits early, the genetic potential of corn is realized, soybeans exhibit nitrogen nodules, and all the small grains serve up bountiful harvests. In the case of cotton, a maximum set of bolls is achieved and the gin turnout in grade translates to a bottom line. A field-ripened early harvest denotes the success of a fertilizer pro-gram because of testing and feeding on a regular schedule. Chandler's many protocols promise better yields, but they also define the numbered steps, without shortcuts, that must be taken to reach that goal. This long learning curve teaches a farmer how to make routine adjustments of the fertilizer program. Every instruction comes with a codicil or sidebar. Protocols call for foliar feeding no less than water management and drip-irrigation according to the layout and the crop, for example. Irrigated crops and dryland crops both invite, indeed demand, leaf or

The Source of Revenue

"It is the flow of energy that governs economics, not the stock exchange or the will of the Congress. In the fullness of time, it is this flow that trashes legislative intent and makes a fool of errant public policy."

— Charles Walters, *Economic Report of the Producers*, 2003.

From his earliest days on a hardscrabble cotton farm, Esper K. Chandler has wondered about more than soil science, crops, and what the plant has to say. The economics under which agriculture operates more than troubled this redheaded boy while he read about Louis Bromfield and Friends of the Land. Later on, when he encountered the work of William A. Albrecht, he inched ever closer to the reality that the equation was rigged.

In *Raw Materials Economics,* a simple statement resonated with Chandler as he went about the business of finding out just what it took to achieve top production with minimum inputs. It argued in effect that if we grow our food crops in the United States, then we have both the food and the money it earns by being produced here. If we buy cheap food from foreign producers, then we may have the food, but the foreigner has our money.

Solar income via chlorophyll dominates our economic system. This means agriculture. To affect a shortfall for agricultural income is to cancel out solar income to the tune of 70 percent of national income. Construction of debt—based banking and exchange does not eliminate the necessity for newly monetized solar income each annum. Stored energy — as in coal and oil — is spent as heat, but solar energy transformed into food energy via the agency of nature's transformer — reflects efficiency that makes all works of man pale into insignificance. Natural humate deposits are increasingly used as biological food in agriculture.

The elemental constant in the statistics is that the malaise can be traced back to what is happening to agriculture. The solar energy revenue has not slackened. Nor have most of the venues that receive and transform it. The culprit is the accounting exploits that allow institutional arrangements to pay agriculture for less than the value received — a national cheap food policy, if you will.

petiole testing. Much the same is true of minerals and micronutrients tested at critical stages.

Soil testing should be done before each cotton crop. Water has to be tested for each source every third year. There is always a caution attached to each protocol. Excellent results have a way of showing up the first year, but a repeat performance is not to be expected without following up. A 200-400 pound yield per acre may be reasonable for cotton, but the cotton grower who uses water properly may be rewarded with an extra bale or two. The cotton protocol offered here is economical, costing only 10-35 pounds of lint per year under the guidance of Texas Plant & Soil Lab. Water, soil, and plant relationships are the lodestones for learning, whatever the crop.

Notes on Peanuts

Peanuts are a specialty crop, especially if they are one of the crops grown on a diversified farm. It is hard to improve on the notes that Chandler and associates hand off to peanut growers: Water + available calcium in a balanced peanut nutrient program are the major limiting factors of yields and quality. All best management practices must be followed to reach the genetic potential of the plant. Aiding optimum nodule-nitrogen production with proper treatments of humic substances and microbes reduces the need for applied nitrogen. Applied nitrogen of the right kind and amount at the right time can improve nut yields and quality. Maximum use of no-cost nitrogen from the air requires calcium in the available form (Ca/H_2O) in the top of the soil's pegging zone. This treatment, as nuts are formed, is the most essential ingredient in yields and quality. Nuts absorb most of their calcium through their skin in the pegging zone. Regular petiole calcium is a poor indicator for peg calcium uptake. The peg stem should be tested separately. Better standards are now being developed for this test.

There are several ways to supply soluble calcium in this zone. Calcium sources and sulfur and/or humic and fulvic acids, lignosulfonates, carboxyls, biologicals, etc., help supply the appropriate calcium.

• Irrigation can leach the soluble calcium below the pegging zone.

• A constant supply is needed through the long pegging and development period.

• Boron aids calcium utilization. Phosphate is needed in constant supply if nuts are to form and develop.

• Phosphate and micronutrients tie up rapidly in highly calcareous soils.

Peanut Field — Seminole, Texas

Gypsum Rate Location Top 0-2" - (ppm) Extraction	0 P-12 Calcium (Ca)		0 P-4 Calcium (Ca)		500 lbs/ac P-13 Calcium (Ca)		300 lbs/ac P-46 Calcium (Ca)		300 lbs/ac P-29 Calcium (Ca)	
	H_2O	CO_2	H_2O	CO_2	H_2O	CO_2	H_2O	CO_2	H_2O	CO_2
Date 8/10/04	24	161	43	225	105	708	27	427	20	145
8/17/04	23	153	30	189	32	297	28	312	23	455
8/24/04	22	128	18	118	14	156	25	181	31	267
8/31/04	17	135	23	159	32	101	29	197	20	176

These West Texas peanuts on new ground attained 6,000 lbs./acre but started to decline with time. Calcium (gypsum) applications to pegging zone leached rapidly during the growing season from the regular overhead irrigation.

• Humic and fulvic acids, lignosulfonates, carboxyls, humates, biologicals, etc., and sulfur aid phosphorus uptake.

• Multi-hormones such as plant growth regulators (IV-types) stimulate the entire plant with better root uptake of nutrients.

• Soil inoculants activate phosphorus uptake.

• Add a regular supply of phosphorus as indicated by petiole tests.

Potash and magnesium are critical at fruiting. They are best applied early and tilled in. The peanut is known as a lazy legume that supposedly prefers to eat off the second table.

Always listen on a biweekly or weekly basis to what the crop has to say. At the end of a crop year, Chandler proposes, it is necessary to test the topsoil to a depth of 12 inches, and the subsoil at 12-24 inches. To sustain top quality production, it's then necessary to test subsoil at 24-36 inches, even 36-48 inches, or to parent material (bedrock), this deeper testing to be done at four or five year intervals.

From Chandler's chair, this much is clear. The cation exchange capacity system — with Albrecht's balances of 65% calcium, 15% magnesium, 5% potash and the trace and micronutrients filling out the equation — was a great step forward in its day. It enabled a sort of prediction as to available calcium. There is no jurisdictional dispute as far as the plant is concerned. Nutrients in the bank account do not assure transfer to the leaf or petiole.

An explanation in terms of peanuts may be in order. The calcium in the pegging zone — where the roots peg once the peanut is formed —

allows more calcium to be taken up through the shell of the peanut than by the roots. The soil and petiole test reveal that calcium may not be taken up to that height, inhibiting the development of the overall crop. Availability of calcium is stalled in the pegging zone. Ca availability on the row and across the side will keep maximum peanut yield if all the pegs make nuts across that row. As the runners go out, pegs go down to do their job, keeping the root healthy and the legume producing nitrogen — based on adequate phosphorus, potassium, magnesium, and a balance of all other nutrients in the soil. The balance, in theory, will deliver. Systematic sprinkle irrigation drives the available calcium below the pegging zone unless more is added or solubilized.

Rather than rejecting the cation exchange capacity system, Chandler chose to build on it, with the laboratory soil and plant analysis as arbiter. The advice given to a field soldier is always the same: It depends on the terrain! In agriculture, it depends on both the terrain and the crop. In the case of peanuts, the peg-zone test is required at least thrice per pegging season. Most crops ask for once-a-year oversight. The plant, once asked, tells which nutrients are available at the right place, at the right time, and in the right amounts weekly. With high value crops, "we ask the plant — first, last and always," adds Chandler as sprinkler irrigation washes down the available calcium.

For crop production, the bottom line always has its say! Therefore, costs have to be spread out and if possible choked to death. Using the best technology available gives the grower a leg up on the market. In the final analysis, it is genetic potential, healthy soil, and plants that write the final entry on the income statement. Quality and shelf life are the final claim to fame.

Notes on Citrus
& Other Tree Crops

Notes on Citrus

"You have to ask the plant when and how much," Chandler adds. This is shorthand for saying that the best way to tell when and how much nitrogen and/or how much of "other nutrients" are required is to crop-log petiole or leaf readouts. Chandler's instructions are always abbreviated, and exhibit a paucity of excess narrative, yet all find space to remind growers that all plants, much like animals, put on the feed-bag each day, not once a season.

Much the same is true of citrus, groves of which have brought fame to the Rio Grande Valley — especially the super sweet Rio Reds. Historically, early harvest delivers profit for this crop. This bland statement depends on controlling production costs, the directions for which make the Chandler instructions worth their weight in gold.

Nutrients have to be made soluble either by soil microorganisms or by the fertilizer fabricator, this unless the soil system is so well governed by the soil's management that a natural nitrogen cycle and a natural carbon cycle are in evidence. There are conditions under which there is enough available phosphorus to last until the next Ice Age. Not so in most of America where humus levels often fall to a half percent or near zero. So, it is soluble nutrients (H_2O/salts of fertilizers, N, P, K, Ca, Mg) which are the first to be taken up by roots if not interfered with by sodium (Na).

In the Rio Grande Valley, there are dry seasons and wet seasons. When rains are abundant, they most certainly leach soluble salt fer-

Quality & High Yield in Citrus Production

Total Nutrient Uptake (lb./ac.)
Fruit Production & Tree Maintenance

Variety	Yield	N	P_2O_5	K_2O	Mg	S
Oranges	25 Ton/ac.	265	55	330	38	28
Grapefruit	25 Ton/ac.	245	51	306	35	26

Many factors affect yields with sufficient and balanced nutrients being a key factor.

The Three Critical Stages of Growth

Prebloom — Stored nutrients must come from leaves and sap. Test late January to early April.

Bloom to Fruit Set — Fruit retention is influenced by nutrient and moisture supply. Test June or July.

Pre-Harvest — At early maturity, fruit size and sugar content are influenced by balanced nutrient levels and moisture supply. Test September to December.

Soil moisture and temperature stress effects can be reduced by balanced nutrition and good management practices — especially potash, copper and manganese.

Soil tilth, humus, biology and energy improves nutrient availability and water use efficiency as well as reducing harmful salt leaching.

Foliar nutrient applications can help supply needed plant food at critical times such as bud swelling and fruit set. Late season applications can speed maturity and harvest is earlier.

Winterize and hasten maturity with foliar sprays. Potash aids winter hardiness, maturity and fruit size. Manganese, copper and zinc are beneficial in late fall foliar applications.

Carbohydrates and humic acid can aid the desired physiological processes.

A leaf analysis is the best way to determine the most beneficial nutrients to help:
1. Set and retain more early fruit
2. Increase yields
3. Improve quality
4. Reduce culls
5. Speed maturity
6. Reduce alternate bearing

Citrus profits come from having early, quality fruit. Balanced nutrition is the first step.

tilizers from the soil. Foliar has been tagged as the salvation for the citrus grove, with small amounts of NPK to be used — perhaps 1-1-1 or 1-2-1 ratio plus other nutrients needed in the Krebs cycle. The foliar route is now wide open since new products have come on line, many of them spawned by the natural/organic movement. In staccato lines, the protocol that Chandler uses to instruct drives home its points with sledge-hammer blows.

Citrus Facts

- First bloom is the first pick at the best price. Late blooms = late pick at a lower price.
- NPK translocates from old leaves to new growth, especially at first flush.
- Other nutrients are not in sufficient amounts to set first blooms, as roots are just breaking dormancy.
- Root activity begins after first leaf growth.
- At first bloom, Ca, B, Zn, Mn, needed for early bloom set, are most often very low in new growth as they do not translocate from old leaves. Therefore these elements must be applied early to set the early blooms for the earlier more profitable crop.
- Apply foliar in February or March just at bud break (or soon after) and at petal fall. It may be best to skip during peak bloom. One application early and one late may be better than one spray.
- Combine several adjuvants: products containing humus forms, multi-hormones, enzymes, carboxyls, biologicals, wetting agents, sugars (molasses), etc.

Citrus Nutrition

Do not allow citrus trees to be subject to moisture stress at early bud break and bloom!

Blooms require balanced nutrients for good pollination and early fruit set. Profits come from the early fruit set that increases early harvest. Foliar Feeding at early bloom and at petal fall aids the retention of the early blooms with an adequate nutrient supply that aids pollination. Cost of spraying at bud break can be reduced by using 30-50 gallons per acre as foliar feeding does not require total immersion as do pesticide sprays. Piggy-backing foliar nutrients with other sprays has no extra application costs. NPK translocates from old leaves to new growth when roots fail to supply adequate amounts — roots are just awaking during early spring flush so other minerals and micronutrients that do not translocate are in short supply.

Calcium (Ca), boron (B), zinc (Zn) and manganese (Mn) are essential for pollination and nutrient translocation for good early fruit set and retention. These do not translocate in the sap of the tree. Even though old leaves may test high, nutrients do not translocate to the new leaves. The younger the leaves the lower they test for Ca and B as it accumulates with age. The longer the leaves stay on the tree the higher the Ca and B accumulates in the tissue.

Early blooms need foliar sprays with small amounts of Ca, B, Mn, Zn and balanced nutrients with adjuvants to aid uptake such as carboxyls, humus, lignosulfonates, multiple/hormones, enzymes, etc. These aid early fruit pollination and set. NPK translocation is usually adequate in early flush. Test early flush with a plant analysis.

At May/June drop, the trees shed the fruit they cannot feed which will be the last bloom if the first blooms (early harvest) are set with proper foliar sprays. Feed properly to make size and quality. Balanced nutrition is essential for good fruit size and early maturity — especially potassium (potash).

Split applications of all nutrients are more effective than large doses. Fertilizer recovery/utilization is greater with adjuvants thus reducing fertilizer costs.

Plants feed and drink every day like animals. Be sure supply is adequate.

Potassium (K) translocates from old leaves to new growth when K levels fall below 0.70% the leaves shed, thus reducing yield and quality potential. Well-nourished and cared for citrus trees can retain all flushes well into the second year thus producing more fruit.

Research at the Lower Rio Grande Citrus Center shows that three foliar sprays of K increased size and earliness. Years of good citrus yields have mined local soils of available potash. The only way to know for sure is to "ask the plant" with a leaf analysis. Test the newest and the oldest leaf on the same limb to measure the health of your trees.

- Rates as little as 5 gallons/acre by air could be highly effective and less expensive.
- Rates of less than 50 gallons/acre by ground are effective and cost less than normal high-volume sprays.
- Double back on a strip to observe better rates, and leave a check strip.
- Only new growing tips need coverage.
- Evaluate about June 1 for location, number and size of fruit.
- Adjust the fertilizer program to match yield potentials and avoid sheepnose.

Chandler's protocol in this case has more details than most others. Chandler has introduced adjuvants of biologicals, hormones, activators, and surfactants. Most of all, the protocol seems to say, "See what you look at."

Chandler takes pardonable pride in the findings his soil and plant consultants turn up by seeing what they look at. There is no "comes first" in the learning department. The act of taking a watermelon petiole test, for instance, can serve up shoots or pointed or rounded leaves, and what is the significance? The same is true for the citrus tree. Here, the point is heart-shaped if all is not well with the world. At the formation stage, drought without irrigation will deliver a heart-shaped leaf, whereas a return of water via irrigation will return the next leaf to normal if adequate boron is in the soil supply. Boron is complicated and misunderstood on citrus because it can accumulate in the leaves to toxic levels and still be unavailable at critical stages of physiological requirement.

It takes both the lab and the field researcher to know when nature reveals her messages and her secrets. Consider sheepnose on grapefruit. At first a calcium or potassium deficiency is suspected, aggravated by high nitrogen. Add to the above a pantheon of traces that almost certainly play a role — for good or ill — and the knowledge still to be discovered both haunting and instructive.

Notes on Oleander

Consider an often unpopular plant, the oleander. Cut down and burned, it delivers a toxic smoke, much like poison ivy. A lot of people near the coast like it as an ornamental shrub because it is tolerates saltwater and drought better than many other plants.

"They have separated a cancer drug out of it," Chandler informs and comments on his work with a grower near Hondo and Medina, Texas, "but they don't know the exact growth objective yet. Do they want to harvest young growth, or mature leaves, or stems up to an inch, or is the whole plant to be harvested?" The drug is extracted out of the tissue.

This is mentioned here to illustrate the range of questions that emerge whenever a consultant intervenes for greater production and a better product.

Notes on Pastures, Turf & Other Grasses

Notes on Pastures

The main crop for most farmers is the pasture. Pasture puts cattle and farm animals into the picture. Pasture consists of real grass, not turfgrass or a manicured golf course. Hooves pound the forage, it's showered with urine, and manure patties pelt the greenery at least nine or ten times a day. Since Chandler has thought of almost everything, I asked him if he had ever participated in the pasture renovation experiments by College Station using dung beetles?

"I would like to say, if I could, that dung beetles are a part of our pasture scene, but the best I can say is that we're working on it," Chandler admits. "One of my problems is an Aggie son who has bought in on the chemical quick-fix, but we're weaning him off the stuff, so he's starting to look at dung beetles. After the extended rainy season of 2007, activity was observed in our working pens."

An expert on dung beetles for the pasture is Walt Davis of Oklahoma. His story has yet to be written. For now, federal assistance has been cut short by a bureaucratic snafu. In fact, using poisons on pastures and dung beetles as an economy measure is an oxymoron, a contradiction that strains the imagination. This lack of insight cancels out the pasture's reason for being, namely pasture grass.

The biggest concern, dung beetles aside, is minerals. Most of the land Chandler farms is deficient in calcium, potash and magnesium. Calcium is the first building-block, and yet coastal Bermudagrass can tolerate a lot of acidity. It has a deep root system, and it mines the subsoil. But when it runs out of nutrients, it goes to pieces. It is subject to winter-kill, disease and competitive species invasion, and spotty performance in general, always because of a mineral shortage.

As a matter of practice, experimental stations merely look at the top six inches of soil. There are still reports afloat that liming does not benefit coastal Bermuda. Few credited the plant for taking its calcium-magnesium lime from the subsoil. It was the development of the in depth soil profile that brought an awareness of the subsoil-lime connection. Texas A&M Ph.D.s Dale Pennington at the soil test lab and extension forage specialist J. Neal Pratt were strong exponents of 0-6 and 6-18 inch sample probes on pasture. They later expanded the concept to include all crops. After them came Vince Haby at Texas A&M's Overton Experiment Station in East Texas. He tried to establish legumes. He had good success sod-seeding alfalfa into Bermudagrass. The results were as spotty as the subsoil minerals, namely calcium, potassium, magnesium and micro-minerals. When non-existent, the results were not even spotty. Where deep minerals were in abundance, alfalfa took hold. He has published this work.

"Back in the 1950s when I was at an LSU experimental station in Homer, Louisiana," Chandler recalls, "we had many agronomists who generally believed it was impossible to grow alfalfa on our deep, sandy coastal soils. We proved that we could, because we brought the mineral content up, especially lime, potash and phosphorus. It was strictly chemical fertilizers. When the same program went out to the public, it failed. If the subsoil didn't provide, the crop faltered. It depended on how exhaustively deep-rooted Bermudagrass had mined the subsoil."

Bermuda prevailed across the South because of its tryst with nitrogen. Chandler recalls an experiment that was supposed to require 400 pounds of nitrogen. It turned out to require 600 pounds for a top yield, as previously noted. In a wet year it was a higher result than in a dry year. Cheap nitrogen delivered more dollars than inputs required. "Here we were, mining our subsoils and declaring a profit. And we should have known better. On a roadside sandbank we were able to make a cut and measure coastal Bermudagrass roots going down 18 feet."

As things were measured during those experimental days, the yield was 8-10 tons per acre. Those same fields circa 2005 hardly got four to five tons per acre. Long before natural/organics delivered a stunning blow to much of the von Liebig heritage, the research that Chandler and his associates accomplished put to bed the idea that "this ground is not good enough for a crop, so it can go to pasture." The profiles revealed with great finality what the problem was and remains. Even so, no standard has emerged for analysis of subsoil. Lime on the topsoil will work its way down. The dandelion is witness to that fact, for which reason its roots go deep to capture calcium and transport it to the surface. In a soluble sulfate form, calcium migrates below quite rapidly. It also depends on good rains.

The question arises: how many fields are in a position to have soluble calcium? Chandler is not aware of even one laboratory — other than Texas Plant & Soil Lab — that tests for soluble calcium. Yet a clear understanding of the nature of soluble calcium is mandatory, topsoil or subsoil. Chandler explains, "The acidification of calcium carbonate comes from sulfur chemically in the sulfate form or biological, with humus feeding the soil food web. You put on your elemental sulfur, and the bacteria oxidize it to sulfuric acid almost instantly if you have the temperature, the moisture, and the fine particle size of sulfur to form gypsum from calcium carbonate. The natural biological action of the organic matter via microbes also acts to solubilize sulfur, calcium, and sodium cations to where they will move. If you have excess sodium, you have to have soluble calcium to exchange on the soil particles with insoluble sodium to make it soluble sodium sulfate. Now it will move with the water." That's the only way to get rid of salt: flush it down through improved soil tilth. Bacterial action converts sodium to a soluble form by solubilizing the calcium and magnesium via exchange, the microbial action finally improving the physical structure of the deep soil tilth.

One anomaly in the pasture is the saline (soluble sodium, mainly) seep. Water goes down and encounters a clay barrier, moves sideways, and then resurfaces. The injection of humates brings back microbes, often defeats the seep, and returns the soil to production. The key is to break up the hardpan. Once broken, the sodium goes south: not east or west, so to speak. The acidification is biological, and not totally dependent on straight sulfur.

Chandler simply tells those who will listen, "See what you look at!" A cow walking across a saline seep often deposits manure. At each dropping, grass springs up as if by magic. That cow is restoring the ecology. This grass production relies on returning humus to the soil. The poverty grass that tries to paint the landscape green may look like grass and feel like grass, but a cow can't thrive on it.

Science has been on a roll for over a century, but it still can't fabricate what the cow does routinely. It strikes Chandler as enigmatic that so many agronomists overlooked the obvious for so many years: that humus as raw manure, crop residue compost, compost tea, or humates governs the biological process. In the final analysis it is the biological process that trumps the simple pie illustration of chemical, biological, and physical soils. It is the biological process that enables restoration of the soil and water penetration.

Here is where natural/organics reached over to shake hands with synthetic chemistry. "We've become so inorganically chemistry-oriented," says Chandler, "that we no longer look at what is going on in nature. The Ph.D. who made that discovery of grass in a saline seep after a manure dropping really kicked open the door of his academia mind, in my opinion."

Happenings such as the above have prompted Chandler to observe, "Why did we wear blinders in reaching for the quick fix? The quickest way was to get a chemical to solve a problem. Agriculture has failed to look at the problem it created." Pastures have suffered more than any other type of land from the pendulum swing between quick fix and total neglect. With native pastures almost gone, and with plenty of cotton and corn ground going back to tame pasture, special procedures must be used to restoring their vitality. "Do a soil profile," is the Chandler way. "It's a case of getting calcium, magnesium, potash, and phosphorus into the equilibrium that old Professor Walter Peevy of Louisiana State promoted and taught me. In converting cotton to pasture, you have to return to a natural balance. This means legumes, the first step in growing a good pasture, whatever the species of grass. It goes back to the Albrecht concept of mineral and micronutrient balance." Legumes need Ca, P, K and Mg in balance.

Chandler can detain the novice for days discussing the grasses that go with special livestock operations. Most of the South thinks in terms of beef cattle, stocker cattle operations in general, with dairies concentrating on the low plains for irrigation water. When a calf is weaned, there is that interval before it reaches the feedlot. The public policy of maintaining grain production has led to animals being put into a feedlot at their weaning weight for the addition of a quick 800 pounds, in effect, creating a ruminant hog.

"During all of this mad dash, those of us in forage production in the South kept pointing out that it was more economical, more healthful for that animal to be fed on grass. It can be demonstrated that when a pound of grain costs 40 cents in the feedlot, the grazer can do it for 20 cents selling grass and sunshine." Chandler insists that the same ratio is still there today.

The Omega 3 to Omega 6 ratio out of the feedlot is one to 18 or 20, with grass one-to-one. Chandler often reflects on the folklore that attends beef production. So the word went out that forage produces yellow fat, whereas forced feeding produces white fat. "In my early days in the stocker-feeder business, cotton and row-crop land were going to forage. Central Texas, Falls County, was the stocker-steer capital of the world. It was a case of row-crop land being returned to winter pasture. Rainfall, of course, is in the winter time in that area. Summers are so dry that crops have a difficult time finishing.

Pastures beckoned, but the trade penalized cattlemen if the steers gained above 600-650 pounds. The feedlots made their money selling feed, not in delivering tasty cuts of meat. They ignored the fact that in 35-60 days on feed, they could finish that meat to grade. No one would buy a 900 pound animal because they wanted one grown their way, the norm being 110-150 days on feed. The standards do change from time to time. Sometimes today an 800 pound animal is accepted.

The livestock industry is controlled by the feed industry, which does not like forage-produced beef. The consumer likes grass-fed beef because of the healthier type of fat. A good sustainable pasture with legumes and different types of grasses delivers a better producing animal, sunshine and grass conferring health. Chandler has been promoting forage-fed animals for some 40 years. He almost steps into the corner occupied by the animal rights people who allege cruelty to animals when kept in confinement, feedlot-style. Clearly, a confined animal is under stress. A suitable pasture-fed animal is healthier itself and for the consumer.

"We used to run several thousand head of stockers. In late spring, the buyers would come. We'd sort through, take the top end off. We hung onto those the buyers wanted to penalize. It was a program of stocking one head per acre, sometimes more with heavy fertilization. Stress produced culls. Grazed under less pressure, the cull animals on the same pasture invited the buyers' attention again a few weeks later. Feeding on matured small grains and native forage, the term scrub now became choice."

Cows are a social animal. The tribe has wallflowers and aggressors. Those who pile the hay in one spot will learn soon enough that some animals thrive while some silently starve. If you roll out a round bale, then every animal gets a spot at the food supply. Chandler recalls how his cow farming family was early in developing Voisin's rotational paddocks. The system has been refined into a procedure called Holistic Resource Management under the guidance of Allan Savory, formerly of the Rhodesia that is now called Zimbabwe. The paddock idea fit in with permanent pastures best. Timely rotation has been and remains the key to both pasture and animal maintenance. Holistic Resource Management has made steady inroads in terms of free-range ranching vs. harvesting sunshine and grass. "We now have a cattle farm, not a cattle ranch," Chandler says. "We grow grass and harvest it with cattle." Rotation has proved itself in terms of animal performance. Chandler cites Betsy Ross of Ross Farm in Granger, Texas and the pastures she refurbished with compost tea. Ross is regarded by many Texans as the "Mother of Compost Tea".

The lesson is: let the grass grow, top it off, and move on. As a consequence, the process builds organic matter. Compost tea is an idea whose time has come. Betsy Ross achieved her maximum gain on pasture treated with compost tea and did not do as well on more abundant, albeit untreated, forage. The difference was a two-pound gain on compost tea, one pound on untreated.

Single factor analysis never tells the whole story. Even conservation tillage can go wrong if the calcium, magnesium, potassium, sodium, and phosphate levels are ignored. Compost tea cannot absolve all the other shortfalls that afflict grass production, but it can work wonders just the same. Often the pasture will have plenty of calcium, but no system for making it soluble. Hardpans can develop even when the soil is not carved with a plow. "You have to look at the whole picture," Chandler says. "The microbes can annihilate hardpan, but first the farmer has to avail himself of what we know. The chant — scientifically sound, replicated, randomized, repeated — holds fast, but it becomes difficult for overstressed farmers simply to get their work done, much less substitute their practices for those of the specialists licensed by law to study and pontificate. Washington believes we're too productive now, so they cut off research resources." This, in light of the fact that the U.S. now imports as much food as it exports, and there is a ballooning trade deficit rather than the surplus produced by agricultural product exports of yesteryear.

The most productive gain per animal, Chandler points out, relies on a short period of concentration, then letting the pasture rest. Winters challenge the system, as do parasites. If the top three or four inches are grazed off the oat plant, for instance, and not overgrazed, then regrowth is faster. It all comes back to utilizing Bermudagrass. For good soil construction, coastal Bermudagrass should be grazed to two to three inches and the roots left alone. The ideal is to fertilize, intensively graze, and get off. The Louisiana State University Experiment Station at Homer, where Chandler was once stationed and helped lay out the rotational pastures, validated this fact with intensified coastal grazing experiments that were widely accepted. Pastures open the gate for a platoon of microbial products that install and sustain life in the soil. This gift from natural/organiculture can be applied to all crops, but it is especially valid for the forage crop. These innovators, along with foliar feeding, have done much to validate what was once lip service to the business of humus, organic matter, the carbon cycle, and the nitrogen cycle. An acre of harvested sunshine outperforms the wizardry of the physics laboratory.

Ken Graff has earned an enviable distinction for raising grass-fed beef on his 7A Ranch near Hondo, Texas, where he also operates a corn maze for tourists. He demonstrated to several commercial hay producers that coastal Bermuda production can be restored to its previous glory, this in Atascosa County, once the prime producer of coastal Bermuda in Texas. Atascosa is now down to a few thousand acres because the soils were mined out and the grass has faded. After thorough soil testing and a program of liming, building humus, and mineral balancing, coupled with the use of biologicals and generous NPK and magnesium, a frost-tolerant first cutting recently yielded 2.5 tons of 15 percent protein. Subsequent cuttings were in the three ton range with soil tilth improving. Soil biological tests were conducted by MK Labs of Fredericksburg, Texas, and special formulations of compost teas were regularly applied. A follow-up precision natural soil test after six months demonstrated the build-up of humus and minerals in the top 6 inches (see table on next page).

John J. Ingalls, a Kansan who served in the U.S. Senate from 1873 to 1891, wrote that "grass is the forgiveness of nature, her constant benediction." Grass is unique in inviting biological sprays, rewarding the user with results only dreamed of in the NPK era. Usually referred to as beneficials, these treatments support life. Chemical herbicides and weed control attempt a sort of birth control, almost always with mixed results. The spin-offs from James Martin and his Medina treatment are legion. Entrepreneurs have enlarged the idea with new microbes, new innovations, and new visions. But the very idea goes back to what Rudolf Steiner, Ehrenfried Pfeiffer, and Martin suggested all along. It

Soil Analysis
Crop: Coastal Hay

| | | | | | | | | * SALT CATIONS - PPM | | | | | | | | | | |
| | | | | | | | | Potassium K | | Sodium Na | | Calcium Ca | | Magnesium Mg | | Ratio | | |
Field	Text.	O.M.	CO3	pH	Salts E.C.	lb per ac NO3	P2O5	H2O	CO2	H2O	CO2	H2O	CO2	H2O	CO2	Na/Ca	Na/Mg	Dates
1 YF 200 0-6	5	0.55	O	6.3	-0.38	52	2	4	11	23	96	29	179	5	24	3	19	02/14/06
2	3+	0.95	M	7.2	0.67	59	2	14	31	59	174	121	769	10	57	1	17	07/18/06
3	3+	0.80	TR	6.4	0.67	23	13	18	35	40	216	47	248	10	39	5	22	12/28/06
4	3	0.85	L	6.9	-0.38	16	38	13	38	25	154	37	247	4	54	4	39	01/10/08
5 YF 200 6-18	5+		O	6.3	-0.38	35	2	3	10	22	100	26	166	4	21	4	25	02/14/06
6	3+		L+	7.1	0.58	35	2	9	19	37	84	102	708	7	38	1	12	07/18/06
7	3+		TR	6.9	0.67	9	9	5	11	47	177	85	310	9	38	2	20	12/28/06
8	5-		VL	7.0	-0.38	3	29	3	11	18	151	18	217	3	29	8	50	01/10/08
9 YFH 0-6	5-	1.10	O	6.4	-0.38	37	8	8	24	26	105	35	294	5	35	3	21	02/14/06
10																		07/18/06
11	4+	1.00	TR	6.6	-0.38	16	31	21	50	38	137	49	422	8	41	3	17	12/28/06
12	4-	1.00	VL	7.3	-0.38	15	44	13	44	15	143	29	337	4	40	5	36	01/10/08
13 YFH 6-18	5+		VL	7.0	-0.38	13	2	5	19	30	125	64	528	11	63	2	11	02/14/06
14																		07/18/06
15	4-		TR	6.6	0.53	7	5	11	22	44	122	56	255	8	37	2	15	12/28/06
16	5-		VL	7.1	-0.38	10	28	3	13	20	164	21	296	2	33	8	82	01/10/08

| | | | | | | | | * SALT CATIONS - PPM | | | | | | | | | | |
| | | | | | | | | Potassium K | | Sodium Na | | Calcium Ca | | Magnesium Mg | | Ratio | | |
Field	Text.	O.M.	CO3	pH	Salts E.C.	lb per ac NO3	P2O5	H2O	CO2	H2O	CO2	H2O	CO2	H2O	CO2	Na/Ca	Na/Mg	Dates
1 YFLE 0-6	5-	1.15	TR	6.7	-0.38	31	6	7	22	26	121	43	253	7	32	3	17	02/14/06
2	5	1.15	VL	6.9	0.58	29	31	24	39	28	56	105	393	6	23	1	9	07/18/06
3	4+	1.00	TR	7.0	1.01	21	20	26	50	46	178	132	634	17	66	1	11	12/28/06
4	4	0.95	L-	6.9	-0.38	11	61	6	34	14	155	17	179	2	31	9	78	01/10/08
5 YFLE 6-18	5+		O	6.4	-0.38	16	1	4	9	24	111	37	216	4	19	3	28	02/14/06
6	5		VL	6.7	-0.38	20	2	10	15	29	58	81	289	5	13	1	12	07/18/06
7	4		M	7.1	0.86	11	2	7	18	55	176	159	750	12	62	1	15	12/28/06
8	5		L-	7.1	-0.38	7	28	2	12	15	209	21	246	2	27	10	105	01/10/08
9 YFLP 0-6	5+	1.00	VL	6.1	-0.38	44	1	11	35	28	123	42	154	6	22	3	21	02/14/06
10																		07/18/06
11	4+	1.55	TR	6.8	1.01	15	18	34	78	37	146	76	470	16	95	2	9	12/28/06
12	5-	0.95	TR	7.1	-0.38	24	33	16	56	13	153	18	217	3	29	9	51	01/10/08
13 YFLP 6-18	6-		VL	6.3	-0.38	22	1	5	11	34	129	47	245	4	26	3	32	02/14/06
14																		07/18/06
15	5		TR	6.7	-0.38	9	9	14	22	47	134	66	312	9	27	2	15	12/28/06
16	5		VL	7.1	-0.38	8	68	3	15	16	166	19	177	2	22	9	83	01/10/08

Six months following compost tea applications we find more humus O.M. build-up and increased minerals within the top six inches.

was Martin, of course, who tapped the ocean — the source of his secret ingredient — for pastures and row crops alike.

Grass cut too short by excessive grazing grows weaker and turns into a thief, forever stealing carbohydrates from the roots. Roots fall into disrepair, often presiding over the death of the grass as lesser species invade. A grass must reach maturity in the fall to store carbohydrates in its roots for a vigorous start the next year, thus warding off competition. Soil life also falls victim to excessive grazing. Compaction replaces aeration. Then the water fails to soak into the soil. Erosion scars the landscape. Carbon dioxide rises into the atmosphere. For stands of grass to flourish, the last cutting must be allowed to attain full maturity in the fall to rebuild and store carbohydrates. This has been well-documented with Johnsongrass, one of the most nutritious of southern grasses, and with the invasion of natives on improved Bermudagrass species.

Chandler kept notes on shortfalls in pasture management over the years, and his laboratory has responded with more answers than farmers have questions. He always starts with the soil audit, the nutrition equation, the stocking formula, and the intelligence that a cow requires about three percent of her body weight in grass each day. The technique called Holistic Resource Management also answered this requirement, arriving in time to save a great deal of range land by governing the stocking rate according to available grass, and by using the cow's own fertilizer to refurbish the pasture, all the while obeying the three percent rule. There are 573 million acres of rangeland in the United States, over a million of which are under Holistic Resource Management.

Mineral balance is essential to efficient production. That is why Esper K. Chandler works in both arenas. Dealing with nature on nature's terms is a grail worth pursuing.

Keeping Grass in Perfect Harmony

Esper K. Chandler relies on good research, and he dismisses company prattle with less than the wave of a hand. Nature has an order, yet some few researchers turn to cooking the books. By following the pathways outlined above, Chandler calls into focus the essential role of water, without which plant growth is less than nothing in the wind. Perhaps half the acreage in North America is in grass, the one crop always in near perfect harmony with life and the seasons.

In the pages that follow, the reader will encounter the drip-irrigation system, the uptake of phosphates as a byproduct of photosynthesis, microorganisms introduced into the plant via the stomata of leaves, and the rules for maximum crops with economical inputs.

"Grass is the forgiveness of Nature, her constant benediction," according to old-time Kansas State Senator John J. Ingalls. What happens to grass seems to happen to row crop agriculture.

Whatever the crop, a return to the miracle of sunlight, photosynthesis, soil minerals, and water asks for attention. This examination of principles is particularly called for when dealing with pastures, turf grasses for baseball grounds, football and soccer fields, golf courses, even polo grounds. Water and plant analysis put turfgrass on the other side of the equals sign. The old eyeball approach probably presides over more athletic injuries than any other cause. Yet plant analysis calibrated together with soil analysis can be used effectively to identify nutrient deficiency as well as existing toxicity. There are soil structure problems related to irrigation water quality and management. The proper application of soil and tissue analysis provide the basis for the natural management necessary to transform a playing field from brick-hard to green mattress consistency with that perfect green appearance. The humus content of soil, the active portion of organic matter, is the foundation of all soil fertility. The natural-extraction method for soil testing is the reason. Carbonic acid in the root zone as duplicated in the laboratory enables identification as to what nutrients in the soil will actually be utilized by the plant. The penchant for painting the turf green with nitrogen and canceling out a healthy perennial crop makes difficult, if not impossible, the use of a soil probe, and even challenges the usefulness of a penetrometer. The rules for the soil test are the same as for the lion's share of crops considered herein. In these and foregoing pages, this narrative has been slow to identify chemicals of organic synthesis often used on athletic fields. Now it is no longer possible to ignore the demand for an ironing board green or a weed-free infield.

In the September, 2002 issue of *Acres U.S.A.*, this writer published devastating data on degenerative metabolic diseases linked to chemical use on the playing field, syndromes such as ALS, better known as Lou Gehrig disease, metabolic disorders, Alzheimer's disease, and the inventory of problems that make nursing homes warehouses for the victims.

All of the above can be avoided, use of 2,4-D included, by use of a nutrient management program. Healthy grass from balanced fertility also uses less water and squeezes out competition.

Addressing Sports Fields & Golf Courses

Special attention has to be directed to ecologically sound turfgrass management, not because pasture forage is exempt from bad science, but because playing field turf management often violates the basic norms of ecological turfgrass fertilization. Chandler met the challenge by telling turf growers for athletic fields to use a soil test and soils that do not offend the cells and protoplasm of players and game officials.

Chandler's protocol calls for an in depth assessment of the status of turfgrass fertility, tilth, and a plant analysis of the playing field, regardless of any nitrogen green exhibited. Such an analysis can be used to identify any nutrient deficiency; because poor grass nutrition enables weed proliferation, which in turn invites, even commands, the heavy use of toxic weed and insect control. Chandler's protocol requires analysis of limitations and problems with irrigation waters, since water is the hydrate part of carbohydrate. The proper application of soil and plant analysis provides the basis for a natural fertility management program. These tests and recommendations are the most thorough and complete in the industry. The carbon dioxide extraction method is a harbinger of protocols used on the full range of Rio Grande Valley, national, and international acres. By mimicking the natural carbonic acid in the root zone, Chandler cancels out the toxic technology that has turned so many athletic careers into a crime that cries out for denunciation.

In his published protocol for turfgrass, Chandler wrote as follows: "Balanced fertility, physiological aids, and adjuvants help each inch of water produce more and better grass. N is a most essential nutrient, about 10 lb. per acre per week is indicated — excess can be harmful. P in smaller amounts is critical for root activity and uptake of all nutrients and water. Uptake is aided by humus products, multi-hormones, microbes, and sugars as energy. P can be supplied to soil, in water, or foliar with all needs together. Minerals and micronutrients are keys to early health and fruit set, growth, and physiology for quality. Soil inoculants and other naturally occurring beneficial soil microorganisms can aid nutrient and water uptake, as well as reduce root pathogens, and protect plants from disease and insects as well as from competitive vegetation.

"You can test turf soils with a soil probe, a turf cutter, or a spade," counsels Chandler. "Slices with a shovel in the last case, take at least four or five random slices from each area." Such a thoroughly mixed sample, dried naturally, provides the raw material with which a more natural turf culture can proceed. Sunlight to photosynthesis, to carbon to water, hence sugars and carbohydrates for the soil and nutrient uptake for plants, all these suggest a clean sweep. Unfortunately, errant technology requires an analysis of what bad science in turf management can deliver, and by implication what that same bad science can deliver to row crops, fruit groves, vineyards, sugarcane, sugar beet fields, and even timber tracts. Soil tilth (condition/structure) for water, air, and root penetration is ignored by most soil tests.

A question that cries out for an answer is posed every time newscasts tell of a sports star cut down by an exotic disease, usually after their playing days are over. Lou Gehrig led the way in the 1930s. The form of lateral sclerosis that terminated his career now bears his name: Lou Gehrig's disease. A few rare studies have linked turfgrass sports to melanoma, other forms of cancer, and of course Alzheimer's. Even fewer studies have noted the increase in disease incidence as parallel to the era of pesticides. Use of chemicals of organic synthesis on soccer fields, baseball diamonds, and football fields is both scandalous and rampant insanity. The worst offenders are managers of golf courses. Almost any golf course manager will tell of five-figure budgets for poisons to banish weeds, maintain turf, and preserve that cosmetic look.

Malcolm Beck has related how soccer and football turf treated with compost reduced high school injuries to near zero, the hardness of ground often delivering broken bones and serious strains. *Ecological Golf Course Management* by Paul D. Sachs and Richard T. Luff answers one of the great needs of sports that rely on turf playing fields. Hardness of ground is not the real problem on the golf course — toxicity is. The real problem is convincing golf course managers that they are either injuring or even debilitating their clients. Some who elect to "go natural/organic" fail because they set out with faulty plans. Players cut low, drop out of sight, and usually end up out of mind.

A study called "Do Pesticides Cause Lymphoma?," produced by the Lymphoma Foundation of America, examined the pesticide problem in depth, mainly by creating databases on farmers, grain-elevator workers, lawn care laborers, but rarely taking on sports except by implication.

Another study, entitled "Proportional Mortality Study of Golf Course Superintendents," was published in the *American Journal of Industrial Medicine* in May, 1996. Golf course superintendents might as well be farming with chemicals as far as exposure is concerned, and the same holds for the workers. Pesticide dusts and years of exposure deliver a high mortality from brain cancer and non-Hodgkins lymphoma. The chances of such a debilitation striking are 2.7 times greater than the incidence in the general public. There are no published data that identify the risk to golfers who walk the turf and do that legendary licking of the pockmarked ball. A University of Iowa study hinted at, but really didn't answer the question why ecological golf course management is the best alternative when chemicals mostly foul the air around the maintenance shed.

"Mortality Study of Pesticide Applicators and Other Employees of a Lawn Care Service Company," published in the *Journal of Occupational and Environmental Medicine* in 1997, merely confirmed that, grass is either ecological or it is toxic, and the players who compete on it are at risk. Here the researchers examined some 36,000 records of employees. Mortality incidence was up slightly, but no real measure was made of degenerative metabolic disease. The creep toward death for younger ages was noted, however.

A few communities are mandating golf course maintenance without the use of pesticides, usually to preserve a safe groundwater supply, including several golf courses on Long Island. Most golf courses go blindly on their way, however, poisoning golfers with toxins that aerosol or transfer their molecules from grass to socks to feet for a mischief run in the bloodstream. Most golf course managers know quite a bit about fertility, turfgrasses, even soils. Unfortunately, they have become locked in a paradigm that showers money on them for inputs and unreasonable demands for instant results. Jim Moore of Environmental Services in Houston, Texas is leading many golf courses to conversion to natural/organic methods.

Conservation Tillage

Conservation tillage is still a bit of a question mark. It slides in and out among the bits of wisdom that Chandler has to impart. It relies, of course, on burndown, glyphosate being the agent. Its wetting agent is volatile, no question. On the soil it becomes food for microbes.

Does Nature Approve of Glyphosate?

There was an old row-crop field to the west, possibly 0.8 acres, outside the lab in Edinburg, Texas. Chandler has this plot of ground that for all practical purposes is waste area. People living near the Monte Cristo Road address object to the rank growth that emerges sooner or later.

The thought came to Chandler to "burn down the weeds," a euphemism for a horror of horrors to organic folk, namely using glyphosate. This led to a reevaluation of that famous burn-down weed-control system sold under more than half a hundred labels. The soil of course was as hard as paving stone. After repeated applications, the soil became mellow and new higher phylum species were showing up. Salty root types gave way to more mellow species. Buffel grass and Rhodes grass, much like Dallas grass, grows in clumps and is salt-and drought-tolerant. Cattle will eat it as they will desert grass. As the glyphosate did its work, weed patterns changed, weeds being an index of what's wrong and what's right with the soil. "The way you make glyphosate more effective is to acidify your water," Chandler advises. "Ammonium sulfate in the water lowers the pH, cutting the rate in half; or use humic acid and other natural based adjuvants."

Merely mentioning a chemical compound for weed control brings the natural/organic grower quivering to attention. "What is the breakdown of the product? What are the dangers and how can the environment cope with them? Does nature approve?" These are the questions that Malcolm Beck asks with reference to glyphosate, urea, hydrogen peroxide, and some few other commercial synthetic products. Missouri's great agronomist, William A. Albrecht, said, "take your choice" when it comes to urea. Urea is a natural product, and it can be synthesized. My book *Fertility from the Ocean Deep* reveals that the ocean has natural hydrogen peroxide, as do the famous healing waters of Lourdes in France. Yet most of the hydrogen peroxide available to dairy farmers and chicken growers is factory made. Does nature approve?

Chandler's view seems to be, "Let the plant (petiole or leaf) analysis decide."

Chandler often brokers out a few study projects, often by asking questions others pick up and try to answer. In any case, something in the cosmos caused Malcolm Beck to plumb the depths of burn-down weed control. The chemical formula for glyphosate reads as follows: $C_3H_8NO_5P.$

It will be noted that glyphosate has carbon, hydrogen, oxygen with phosphorus and nitrogen — no chlorine — in its formula. The issue that frightens laymen and professionals alike is toxicity. Further, the chemical handbooks nowadays downplay LD_{50} — the measure of what constitutes a lethal dose — and submit well-crafted rhetoric to suggest that toxicity isn't really that toxic after all. In this case, the manuals may be right — for soil, that is. We'll defer danger to the spray operator for now.

Glyphosate is made of the components of sugar: carbon, hydrogen, and oxygen, plus phosphorus and nitrogen. Here we have an energy source (a carbohydrate) plus two minerals that a plant needs. Glyphosate does not linger in the soil. Microbes and earthworms eat it up rapidly. What makes it work is the surfactant which is part of the preparation. The surfactant presides over penetration into the plant via the green leaf. When the surfactant carries extra fertilizer into the cellular structure, the plant becomes so loaded with pure food in the vascular system that the overload kills the plant.

"In Texas, glyphosate is gone in a few days," Chandler decided after study and consultation with Malcolm Beck. Farther north, where weather closes down microbial activity most of the year, glyphosate may survive in the soil longer.

"There is no breakdown product," Beck and Chandler assure farmers. In the soil, glyphosate awaits a temperature of 55 F for breakdown activity. In the human lung, intake both damages and kills. Professors and chemical professionals assure one and all that they can drink the stuff with impunity — impunity if the charge goes to the stomach. If it goes down the windpipe, it becomes a one-way trip to the underground bunker. Much mist spraying allows the aerosol to be inhaled — the real danger is the wetting agent used according to Beck's sources. Application should be in droplets.

The debate has become a standoff. The plowman school expects to annihilate weeds with tillage that destroys mycorrhizae, and the glyphosate school holds to judicious use with the result that when fungi no longer get carbohydrates from the weeds, they emit spores.

That's the short version, but it leaves unanswered questions that nag the cause-and-effect wonderment. Well, wonder no more.

As scientist Jim Cumming of London, Ontario puts it, "The glyphosate formulation can be quite nasty." The original Roundup contained a wetting agent, a trade secret. Glyphosate itself has proved deleterious to amphibians, frogs and toads. Further, says Cumming, "The active chemical in glyphosate does damage to the human liver." This clear evidence was discovered in South America. It acts like estrogen and feminizes, affecting the fertility of males. The route of entry is the skin or diet. As more and more glyphosate-ready crops come on the market, the government agencies raise the tolerance levels for the glyphosate contaminant in food. The objective is to allow the consumer to make it to the door, but not to erase sales. In the original genetically modified crops, an additional gene was added. The plant was made insensitive to glyphosate. That gene never worked very well. In order to increase the tolerance, additional copies were added. That meant that the manufacturer could pump more glyphosate onto food. When farm animals consume such crops, they get their fix of glyphosate, and this is passed on to human beings. Frogs and toads seem to be the proverbial canaries in the coal mine. As this feminizing agent works its wonders, it changes the human liver and delivers mischief still to be plumbed. It must be remembered that such formulas contain unlisted products, perhaps a chemical to cause less aerosoling and more coagulation. Acrylamide is such a contaminant. It is a plastic. It is often used in soil via drip, flood, or spray irrigation to prevent leaching. It delivers a hammer blow to food. It is rich in nitrogen, much of which ends up in the ambient air. High rates of nitrates in drinking water add to all concerns.

That's the long version.

Chandler looks over his former rank weed test plot, "I marvel at how little we understand nature's way."

Defending
Natural/Organics

The university professionals had their settled body of knowledge reaching back to the days of Justus von Liebig. They claimed that salt fertilizers trumped biotic life and even physical aspects of soil management. There were few who could answer. There were hardly any professional people who could discuss humus, organic matter, soil life — except to render lip service. But we are what we eat and we are what our farms produce. The steady march of degenerative metabolic disease does not speak well for our prevailing form of agriculture. The course of human events has ordered change, yet change has been kept an arm's length away. Now it seems that living life is again becoming the ultimate objective of select farm leaders. Now we can begin to construct an understanding of the ecology of the soil. Nature knows no time limits, is ever forgiving and is able to repair herself from our ignoble treatment, in spite of our capacity for destruction.

The confusion touched off by Rodale's adoption of the word "organic" instead of "biologically correct" has never fully subsided. Organic means carbon-based. The soil mineral becomes organic when it is taken up by the plant, having been acted upon by microorganisms, the size of almost all entry being the ion. As with most terms in the English language, organic has other meanings. Thus, "natural" — as opposed to "synthetic" — means organic growth. But semantic jousting cannot settle the issue.

Appendix A

K. Chandler —
The Making
of an Agronomist

The Louisiana Homestead

K. Chandler was born on a former sharecropper's farm in 1926. The homestead where Chandler grew up and gained some of the foundation for his natural insight was 180 acres, hill country later given over to pine reforestation. The nearest town was Dubberly, Louisiana. His grandfather on his mother's side was part Cherokee. Pioneer traffic down the Appalachians and across the South saw to it that the races shared their genes while settling the land. Grandpa worked in the copper mines before migrating West in the 1880s. The work available was timber harvesting and sharecropping, usually a cotton crop that was no longer king, perhaps not even a worthy pretender. Grandpa worked at blacksmithing and ginning for off-season income. He and his first-born son also worked with mule teams to help build railroads, roads, and pipelines. The usual procedure was to harvest the timber, clean off the stumps, mine out the land with crops, then move on.

Chandler grew up with cotton as a cash crop. Many of those acres were poverty acres. Timber harvest was usually followed by erosion. Ten years seemed to exhaust the soil. Most farmers moved on, secure in the knowledge that they were "still young enough to wear out another farm." Off-farm work — often oil field work — meant the difference between relative affluence and subsistence farming. Four families lived off of those 180 acres under tillage during those Depression years. It was a life of chop the cotton, milk the cows, sell the butter, and bless the hogs, chickens, and dogs with skim milk. It was the daily work of stacking manure and then hauling manure from static compost piles to the field for fertility. Even the least trained farmers knew enough to use peanuts, velvet beans and soybeans to hold the soil, to live on, and to swap for vegetables and specialties.

After a tornado claimed their house in Cotton Valley, Louisiana, Chandler helped rebuild, while working as a roustabout in the oil fields, finally settling for employment on a plantation near Gilliam, Louisiana, with the most progressive farmer in the area, Dan Logan.

It was here that the old science and the new science clashed. The great professors of the 1920s and 1930s, William A. Albrecht of Missouri, Firman Bear of Ohio, all those who met at Louis Bromfield's Malabar Farm as Friends of the Land, had defined the direction for agriculture, but their work remained largely unknown. A neighbor, Ruben Douglas operated his All's Well Farm on these principles for Chandler's inquisitive attention.

Their system was based on cation exchange capacity, base exchange saturation, and bringing calcium, magnesium, sodium, and potassium into equilibrium for the next season's crop. The anions also were accounted for, and due attention was paid to organic matter, humus, and the billions of unpaid microbial workers upon which fertile soil depended. Albert Carter Savage's *Mineralization, Will it Reach Us in Time* may have caught the attention of Winston Churchill, but many of those lessons died aborning.

Higher Education

World War II was the event that separated the rigors of the 1930s from the innovative career more than a few call the Chandler Effect. Navy service gave him access to the G.I. Bill, which made it possible to study at Louisiana State University (LSU), where he took a Bachelor of Science in general agriculture. By that time, great professors of the Midland colleges had advanced biologically correct agriculture to a take-off posture that was not to be extinguished until the arrival of Poison Control Centers in 1949.

In the 1950s, Professor M.B. Sturgis assigned Chandler's graduate agronomy class the task of evaluating the idea that J.I. Rodale was promoting, which was based on the work of Sir Albert Howard. They found the premises to be correct. The first line of defense against bacterial, fungal, and insect attack was healthy soil, ample humus, and microorganisms, on the march ever ready to make mineral uptake available for plant roots. It was the atmosphere of the times that halted progress in that direction. Natural/organics seemed to counsel a retreat to mule power, to the resources of a virgin land that no longer existed.

A career as county agent beckoned. Chandler was the first recipient of a rural sociology B.S. degree from LSU, a consequence of reorganization that saw his degree moved from the school's Liberal Arts locus to the College of Agriculture, with many agricultural science courses as electives.

Albrecht, Professor of Soils

William A. Albrecht's stamp on sustainable agriculture is as valid today as it was when it first gained access to peer-reviewed literature. And he knew that he had to begin with calcium. "Your acid clay is nothing more than one that doesn't have the positive ions on it — hydrogen, calcium, potassium, magnesium, sodium, and the trace elements. I've got to have 65 percent of that clay's capacity loaded with calcium, 15 percent with magnesium. I've got to have four times as much calcium as magnesium. You see why we ought to lime the soil? We ought to lime it to get it up to where it feeds the plant calcium, not to fight acidity."

Albrecht expressed these findings in a formal paper given at the International Society of Soil Science on the day that Hitler moved into Poland in 1939. He presented formal papers in Russia and Australia as well, but in a manner of speaking, Albrecht did his best work at home. He stood alone against the fertilizer industry, for instance, even when speaking to its face. "Fertilizers are made soluble, but it's a damn fool idea! They should be insoluble but available. Most of our botany is solution botany. When we farm as solution botany, the first rain takes out the nutrients."

Bromfield & Malabar Farm

Albrecht met often and at length with Friends of the Land, and became a regular guest at Louis Bromfield's Malabar Farm. He liked Bromfield, served as his advisor, and watched with keen interest progress on the farm where Bromfield had set out to prove that natural farming would work.

Bromfield had seen most of the world's abused acres. Indeed, for some twenty-five years he was known as a citizen of the world. He lived in India and Europe, and he wrote novels — *The Green Bay Tree, The Rains Came, Mrs. Parkington, Early Autumn.* The latter won him a Pulitzer Prize.

During the 1930s the world came unglued at the seams. Bromfield's harsh, critical tone seemed to be offset by knowing he could do nothing about it. And so, in 1939, he came "home," to a place called Pleasant Valley in Ohio. Here he found a situation he could do something about. Land had been abused. Bad practices and poor cultivation had filled the territory with erosion gullies and poverty. Bromfield picked up one farm after another. Using what we now call eco-farming principles, organi-culture, and sustainable agriculture, he produced abundant crops.

Bromfield did more. He studied the works of William A. Albrecht, and in a practical arena started to prove that insect damage and disease could be controlled with humus, plant nutrition, and sound soil management. In *From My Experience,* published in 1955, he cited his finding that "insects showed an aversion to all plants grown in good, balanced, living, productive soil."

At the time public policy was trying to cast agriculture into the role of an industrial procedure. Bromfield spoke out, calling for preservation of the nation's topsoil, and he proved out the role of the individual farmer on the Muskingum River watershed, where small dams and soil management practices showed their worth as opposed to big dams and naked hills. At one time the big soup companies sent their agents to Bromfield's Malabar Farms, and for years agriculture's leading intellectuals saw the road it ought to travel. Then something happened. Before many years had gone by, the nation's schools and extension agents — acting as one — touted toxic technology. The Malabar experience was eclipsed.

Chandler & Albrecht

Chandler read about such things as a youngster because the press had no incentive to suppress such reports. Others less flamboyant than the famous novelist also had their say. The sheer desperation of animals fed on deficient forage emerged time and time again in William A. Albrecht's writings. "Our dietary essential minerals are taken as organo-inorganic compounds. We are not mineral eaters. Neither are the animals. When any of them take to the mineral box, isn't it an act of desperation?" On another occasion, Albrecht came right to the point. "Cows eat soil or chew bones when ill with acetonemia, pregnancy troubles, or deficiency ailments. Hogs root only in the immediate post-winter period after being confined to our provision for them. That behavior suggests past deficiencies to be quickly remedied by desperate digging in the earth." Even in retirement, Albrecht worked overtime at schooling people. He corresponded tirelessly. People with minimum training could count on this quiet gentleman to explain complicated chemistry in terms they could understand.

Albrecht's visitors were mainly farmers, consultants, and students, but well-known people of vision also came by to pay homage. One day, quite unannounced, this writer drove down to Columbia, Missouri with Eddie Albert, the movie and TV personality. Albrecht gently excused himself from a recording session with a university archivist and gave Albert both barrels: an hour-long lecture on the Pottenger cat studies, and another on the miracle of the Ozarks. I had handed him my book, *Unforgiven,* both to defend myself and to prove that economics could be a science, hinting slightly that true science was being distorted by the makers of salt fertilizers and toxic genetic chemicals. Albrecht responded, "I have read your book. You do good work. And you're on track."

Those last four words meant more to me than a medal etched in gold. There was no reason why this gentle old man could not have lived longer. His health was fair to good, and he knew how to take care of himself. "I know how they'll get me," he confided. "One day they'll get me out to one of those restaurants with all that synthetic food and preservatives, and I'll suffer an attack. I won't be able to get to my bathroom. They'll take me to a hospital, and they won't know what to do. I'll be a goner."

It happened that way. A group of colleagues ordered Bill Albrecht to a restaurant to honor him. In the convivial atmosphere of the hour, he made an error. He ate chemical-laced fare that triggered a metabolic problem he had been living with for years. They took him to a hospital. In a few days he belonged to the ages.

One day, not long before he passed away, Dr. Albrecht told me where I could find still more news for *Acres U.S.A.* "You go see Gene Poirot down at Garden City, Missouri. He is one smart old cookie," Albrecht said. I did that — and when the time came to publish a fair selection of Albrecht's papers in book form, I asked Eugene M. Poirot to join Dr. Granville Knight in writing a foreword. Poirot had first met Dr. Albrecht in 1926.

"We began research for making artificial manure out of straw in order to have a bit of organic fertilizer. This, we hoped, would encourage the growth of nitrogen-fixing bacteria on sweet clover in soil where the organic level was too low to support the bacterial life in then available laboratory cultures of inoculation materials. Later I was granted two patents on the process which was made available to farmers by the Capper Foundation of Kansas," wrote Poirot.

The Albrecht Papers told this story. They told more, but not all. For it is a fact that Albrecht's findings enabled the nutrient-deficient Ozarks to produce good calf crops so that now Missouri is second only to Texas in calf production. William A. Albrecht lived his life as a scientist, a writer, and a speaker, always serving tirelessly as an interpreter of scientific truth so that inquiring minds could use this knowledge in service to the human race.

Chandler Arrives in Michigan

It was not the knowledge that Dan Logan sought, although he fervently followed corn with hairy vetch plowed under for the next two cotton crops, which were yield breakers of his time. He taught Chandler that cotton could be a good income crop. His forté was modern scientific agriculture based on salt fertilizers and toxic rescue chemistry. Anything new that came down the pike found a willing customer in Dan Logan. Dan earned a niche in the history of cotton production.

When Dow came up with a yellow herbicide as a pre-emergent, Logan went to Midland, Michigan to offer his farm as an experimental showcase and sales operation. He wasn't interested in peer review, only in practical application. About the time recent graduate Chandler joined Logan as a tractor driver, herbicides were coming to the farm.

The first 2,4-D used on a farm in Louisiana was applied by greenhorn Chandler in that commercial setting, this on corn. Double-cross hybrid seed corn seemed to respond on river-bottom land. It not only served up crops, but decorated them with bindweed. Usually the John Deere corn picker tied up more time pulling vines than it did picking corn. So when 2,4-D came along, it was plugged in immediately.

"There was this big imposing former county agent from Texas with his white cowboy hat and white Cadillac convertible," Chandler now recalls. "He was with Sherwin-Williams, the first people to tout 2,4-D. The county agent commanded attention. 'We're going to put it on this corn. You rig up that detasseling machine.' It was a three-legged monster with a water-cooled engine and a raised platform. 'We'll put barrels on that and spray this corn.'" One day, while Chandler was crossing a field drain, the apparatus flipped over, and he had to come up with some nimble acrobatics, thanks to tumbling classes in high school, as a barrel chased him down the field.

Blessed by expert direction, Chandler put out 2,4-D. It turned out to be 2,4-D ester. "That's how ignorant we were," Chandler admitted.

Cotton was being planted at the same time that 2,4-D was applied to the corn. About a month later, a foggy, misty morning arrived. The 2,4-D volatilized. It drifted to a neighbor. "You could go two and a half miles up that river bottom trailing the path of the wind," Chandler observed. "It killed a few plants across the turn-row, then it meandered across those fields. The old fellow operating the neighboring farm was a retired army finance captain. Chandler never forgot Captain Robbie Dole. "The damn fool," blustered the Captain. "If Dan Logan farmed the way he knows how to, instead of chasing every damn fool idea, he'd do all right." The average yield was known, the loss was calculable, and Dan Logan offered to make good any losses, but the Captain refused to accept compensation for the damage, as there was an unacknowledged increase in historical yield. "I know Dan means well," came the laconic reply, "but he's still a damn fool."

The idea that weed and insect control were seated in fertility management had a few supporters in those days, but more sufferers. Modern farmers were looking for foo-foo dust, for miracle cures, not for information on natural/organic farming. The observable fact was that 2,4-D had taken care of the bindweed for that year.

Chandler was in charge of 48 sharecroppers. These families earned 40% of the crop by hand-hoeing weeds and grass and picking the cotton. They drew weekly "furnishing" to eat on during off-seasons and $1.50 per day or per hundred pounds of cotton picked against their yearend settlement before Christmas. When their share crop was tended, they could work in day labor gangs at $2.50 paid weekly to non-sharecroppers. Almost all families had a cash settlement after working all year. His best family earned over $3,000 cash at years end. Chandler was drawing $150 a month plus two meals a day and a room in the Bug House, a camp house that Logan had built for the Logan Seed Company's four entomology teams, the first organized cotton-scouting teams in the nation. That year, he earned a $300 bonus as the assistant farm manager.

By fall of 1949, hand labor was giving way to the Rust cotton picker and other innovations. It was mounted on an Allis-Chalmers machine. These pickers failed on fields damaged by hurricane wind and rain. Logan turned to the Bracero Program, and Chandler was exposed to the hard-working migrant workers. Folks were brought from as far away as the Brazos River bottom near College Station, Texas. Later, Chandler, as manager of the Sneed Estate near Calvert, Texas heard the locals talking about that year of handpicking at Gilliam, Louisiana.

On Logan's farm, Chandler met up with Logan's concept of rotating corn with cotton. Cotton behind corn called for hairy vetch. He would fly hairy vetch seed onto his cotton acreage in the fall before the cotton was picked. The vetch would come up under that cotton and grow under the shredded stalks. With hairy vetch, a new record was set: 2.5 bales per acre across 100 acres. Corn was followed by vetch plowed under with the stalks. When a hurricane struck and the gins couldn't clean the trash, Logan helped capitalize a new gin replete with lint cleaners.

How was it possible to criticize anhydrous ammonia and 2,4-D when it helped deliver a record crop? Quite late, it came to Chandler that 2,4-D was a hormone, a growth hormone. At full strength, it causes a plant to grow itself to death. In diluted amounts, it hoped to substitute for the natural strength of nature's own hormones. It was a synthetic gibberellin, and it made a better cotton crop. It eluded the deathblow of a weed killer, but it left mischief in its wake. But even after experimentation, it proved difficult if not impossible to get the right dosage. This remains the problem today with synthetic hormones. "We're trying to duplicate nature," Chandler summarizes, "but we can't quite find the balance." That's why the seaweeds with their auxins, cytokinins, and gibberellins are winning out. Chandler looks back on the Logan era with abject disbelief and wonderment at his early learning curve of chemicals pushing out natural methods.

Anhydrous ammonia also was given room and board in Logan's fields. An application of anhydrous would literally annihilate the soil's microorganisms, but recovery was so swift due to high soil organic matter from crop rotations and legumes and nitrogen supplying the missing energy source, that this defect was actually seen as a plus. The early literature on this source of cheap nitrogen made farm paper headlines.

Still, rotations, balanced nutrition, cover crops, and practices now treasured by sustainable agriculture hung on as if to recapture the values. The rhythm of one year of corn and two years of cotton enabled top production status in the local parishes.

The Beck Influence

"After I came home from Naval Aviation Cadet training during World War II, I went back to college and majored in agriculture. The truisms published back then stayed on with me. Even though college training guided me into chemical agriculture, I never left the basics of soil fertility. To produce — whether you're coming from a natural or a chemical point of view — you have to have the soil organisms in balance in order to achieve effective production. This training via a few publications allowed me to work with natural/organic farmers for a number of years. When Malcolm Beck first started, we visited his farm with chemical suppliers and soil testers in tow. He had just moved onto that farm."

The Korean War

Chandler took off two years for the Korean War, where he served as a combat infantry officer. His three-month all-expenses-paid vacation to the Tokyo Army Hospital, thanks to an ambushing Chinese burp gun, earned him a Purple Heart and exposed him to extensive Communist propaganda. He learned Marxist theory from literature that his roommate received from his UCLA Communist Front student wife. He returned to the farm, then went back to Louisiana State University for a Master's Degree.

Back to Louisiana

It was now a world of the Aiken Bill, sliding parity, cotton allotments, and agriculture with the status of a public utility. It was believed that by controlling storable crop values from 82% of the harvested acres, all of agriculture could be controlled. In the late 1940s, the movement toward modern farm technology soon acquired the force of convention. Farm Journal said there were too many farmers. High tech became the buzzword of the era.

If you take the time to visit a museum or go behind the hog pen of a dilapidated Southern farm, you might encounter a junkyard relic of the original Rust cotton picker. Pushed to describe it, one would say it resembled a tractor with an outsized basket up front. Legend and researched historical fact has it that the Mississippi town of Clarksdale was the site of an event on October 2, 1944 that changed cotton farming forever, and at the same time set off a migration out of the South that may still have yet to run its course. Some 20,000-30,000 people invaded the Hepson farm for a demonstration that reverberated in Washington, caused revision of Roosevelt's farm policy, and finally savaged the very idea of farm parity.

John and Mack Rust were Texas populists who spent most of four decades inventing a cotton picker. The news moved across the South like ink on a Rorschach blotter test. The Hepson meeting hoped to deliver the word in an orderly manner before rubberneck visitors annihilated the current crop and the fragile terrain. In fact, the Rusts had been demonstrating their picker for a decade or more before they became bankrupt. Their machine could pick cotton, but it could not be produced by lone mechanics in a machine shop. Not being able to pay their taxes, the Rust operation was dissolved. Logan was an early investor in, and promoter of the Rust machines.

New innovators picked up on the idea. Allis-Chalmers and finally International Harvester took over. Cotton was rotting in the field for want of labor. Al Hepson felt this shortage more than most, so he adopted efficiency as his working mantra. He became a scientific farmer. He managed his fields, used bookkeeping to replace seat-of-the-pants records, and invited rescue chemistry to the table as well as rated bagged fertilizers.

The cotton picker had mechanical fingers that stripped off the cotton so a vacuum could suck it up for transport to the collection basket. A good field hand could pick 20 pounds of cotton in an hour. A mechanical picker could collect 1,000 pounds an hour. Hepson's machine could pick all the cotton on the plantation. The cost for hand labor was $39.41 per acre. The machine cost $5.26.

By the end of the Korean War, folklore subsistence farming was in retreat. Perfectly valid early lessons passed on from land-grant colleges were swept under the rug. Industry took over much of the role of intellectual advisors. The great transition made itself felt in the university. Double-cross hybrid corn started in the land-grant college system. Chandler's boss, Dan Logan, was president of the Louisiana State Hybrid Corn Grower's Association. "He led the chemical revolution," Chandler observed, "and we followed."

Chandler returned to LSU in 1953 with Logan's encouragement to start graduate school under the tutelage of Dr. William H. Willis, a very demanding professor of exactness. His major was soil microbiology in soil fertility, and that was when he became exposed to William A. Albrecht, the Missouri University Ph.D., and J.I. Rodale. One of the class projects was to research what the natural/organic folks were doing. Was organiculture fact or fiction? "Our conclusion was that it was based on good scientific elements, but that it represented rolling the clock back at a time when the chemical revolution was sweeping everything in sight."

An advanced soil fertility course taught by Bert Driskill, a recent Ph.D. from Penn State, introduced the field quick-test petiole kit. The other professors derided Driskill for wasting time, but the idea caught on with his students, and Bert made the kits and supplies until well after his retirement back home in Alabama. Dr. Nevin Morgan was an early practitioner, and taught Chandler practical applications. William Patrick was a fellow graduate student who played a key role for years in Chandler's professional learning. Patrick later gained fame as a rice and coastal soils expert.

Phillips 66

Chandler was recruited out of LSU before he finished his Ph.D. work by the Phillips 66 Petroleum Company, then one of the world's largest nitrogen producers. It was a prestige job in the fertilizer industry. The company had a merchandising program. It followed the traditional producer-manufacturer-dealer-farm customer institutional arrangement of the time. Phillips hired agronomists with an education and product in mind, this to introduce their products to farmers via FFA teachers and Extension 4-H clubs and to work with research and Extension.

Chandler's attraction to Phillips went back to boyhood days. Waite Phillips gave his extensive New Mexico properties to the Boy Scouts of America, a program that saw Chandler spend ample time at Philmont, the Scout ranch near Cimarron, New Mexico. The Boy Scout interest followed Chandler to college, where he joined Alpha Phi Omega, a scouting service fraternity. One of the way stations to farm consultation led to association with H. Roe Bartle, the colorful Scout executive who became mayor of Kansas City and for years brought an annual Scout Jamboree to that "up-to-date" city.

The training was more than a career aside. In scouting, Chandler learned how committees functioned, how organizations ran, how organization at the grass-roots level made that level the effective level for any project. Ten years after a Philmont summer, Chandler met up with the agronomists who made Phillips' agriculture tick. This meeting gave him an epiphany of sorts. He wanted to know what anhydrous ammonia was doing to the soil. The conventional wisdom was that if a little did a lot of good, put on more! Buffered by rotation and legumes, anhydrous became a magic word. The stuff is still sold today because it remains the cheapest source of nitrogen, but high transportation, security and insurance costs are drastically reducing its direct-to-soil use.

Andrews Answers
the Anhydrous Ammonia Questions

W.B. Andrews, Ph.D. of Mississippi State, was a pioneer of anhydrous ammonia research. Anhydrous-fed soil often delivered top yields. That was the way of things when soils with high organic matter were starved for nitrogen. Record crop yields lasted some 15 to 20 years. It took that long to finish off the soil humus. W.B. Andrews did the microbe counts and codified the full range of observations on populations knocked down and on populations recovered. He saw benefits slowly eroding away like most flash-in-the-pan miracles erode. It meant that excess carbon with no nitrogen to decompose it led to spot sterilization with anhydrous ammonia, then to reinvasion from the periphery. Yes, fanning the microbial fires made headline yields for an era. As carbon was burned away, organic matter replacement became the problem.

It was to take decades for a new paradigm to emerge. The point Chandler makes when woolgathering about the cutting edge stuff that came his way in those days is that crosscurrents were always afloat, turning the tide this way and that.

Bermudagrass & Mining the Soil

At one point, LSU recruited Chandler to troubleshoot at an Experimental Station in northwest Louisiana. The station had been recommending 400 pounds of actual nitrogen per acre per year on coastal Bermudagrass. The old professors told Chandler to "go up there and straighten those guys out. There ain't no way you can use 400 pounds of nitrogen on Bermudagrass, or on any crop." After a very wet year, a professor asked Chandler what he'd found out. The answer, confirmed in a database, stunned the schoolman. Deep-rooted plants, sandy soil, and heavy rainfall: all said "600 pounds of N per acre" on Bermudagrass was profitable.

There was a road cut at that farm. Chandler and co-workers chased Bermudagrass roots down 18 feet. "We were mining the subsoil and didn't know it," he recalled. "On that station we weren't getting a response to lime though the pH was dropping radically. We were using ammonium nitrate, nitrate of soda, urea, ammonium sulfate, and anhydrous ammonia sources on 7 x 14 plots and hay production fields, 30,000-40,000 bales per year."

The point: 600 pounds was economical "if you didn't look to the future," Chandler summarized. The man in charge of the University Extension Soil Fertility Program was Walter J. Peevy, Ph.D. Even in the 1950s, Peevy believed in rotations. He considered liming not only to correct pH but also to load the soil colloid with the prince of nutrients. He counseled building up the phosphate level. Potash commanded attention and legumes belonged in the rotation to make the natural nitrogen cycle work. Peevy was also a pasture man. His goal, and the goal he passed along to Chandler, was farmland recovery. He wasn't tied to high-test and high-rated salt fertilizers.

The business of mining the subsoil while applying the very same nutrients to the topsoil and waltzing into uneconomic procedures started troubling honest researchers shortly after the new technology and new economics left the launching pad.

NPFI Recruitment

The National Plant Food Institute (NPFI) was the next to recruit Chandler for service in the southeastern and southwestern states. The American Fertilizer Trade Association had already merged with the Plant Food Institute, so now there were two branches: one dedicated to research and development in order to expand the knowledge needed for wise and efficient use of fertilizers based on soil testing, and the other a lobby for the trade.

Russell Coleman, Ph.D., the Mississippi State Director of the Experimental Station, and Samuel Tisdale, Ph.D., out of Auburn, Georgia and North Carolina, headed the new joint organization. The objective was the one that most of agriculture seeks today: a good, sound program for the farmer. Extension, experimental stations, lending institutions, the media, and the industry stood ready to rescue agriculture from most of its drudgery and all of its folklore with intensified soil fertility programs.

Looking back, Chandler counted each new experience as the foundation for learning — tied first to the sustainable objective, then to the motives of industry. Lessons from both founts of knowledge were hard to refute. Those programs doubled and tripled farm income. But by the early 1960s a dissenting voice brought rain down on the fertilizer parade.

Silent Spring Starts Changes

Silent Spring was an instant best seller. In it Rachel Carson editorialized, "We are rightly appalled by the genetic effects of radiation. How then can we be so indifferent to the same effect produced by farm chemicals used widely in the environment?" It was a whisper in the wind, a taunt from the wings that mattered little to farmers who spat out the word "organic" like a foul persimmon.

Chandler knew something was amiss. His career came to involve seeking out research that was being ignored by peer review journals. New insights on calcium, phosphate, and potassium came to the fore. Humus studies joined the "values" studies. "Robert Petit of Texas A&M did the early work, but he was kept in the desert," Chandler recalls, "thus defining the metes and bounds of his career. He and Dr. Flake Fisher were two of the first to suggest putting humates in chemical fertilizers, a sales success for the Best Fertilizer Company."

The road back to valid values became compressed in the extreme when officials with the Association American of Plant Food Control pushed model laws for statewide adoption. That organization, under pressure from its powerful industry members, came up with the idea that it took 20 to 24 total units of NPK to constitute a complete fertilizer.

Best Fertilizers

Chandler joined the fertilizer fray when NPFI terminated its research and development. "I really wanted to go back to LSU to finish my Ph.D. Arrangements were already made," he admitted, "but the phone rang one Saturday. Ken Winborn was on the line. Ken was the most successful independent fertilizer man in the country. No formal education, but one hell of a businessman who merchandized the agronomic benefits of sulfur in his products. I had competed with him selling ammonium sulfate. He brought blending fertilizers to the southwestern states and was also the last large producer of 0-20-0 in the nation. My thoughts were that he was the last person I wanted to work for, and Houston was the last place I wanted my family to live. My wife Bettie reminded me that we would listen to all offers before making up our minds. That turned out to be the wisest counsel from a financial and business standpoint ever heeded. Those ten years in a fast-developing industry with Ken, Lowell Berry and others would make a good adventure novel."

The rest of the story plays like a good drama. Winborn was the executive vice-president of Best Fertilizers of Texas, which was owned by Lowell Berry, a self-made fertilizer industry leader who started his business by making a 0-20-0-based lawn fertilizer in 1933 in his garage. As innovators, they brought blended fertilizers to the Southwest to replace the homogenous, formulated, regulated grades. Berry insisted on hiring agronomists as salesmen, who used soil and plant analysis to create crop formulas fit for each individual field in order to render the farmer more profits. Consequently, the company had better profits. Berry used a profit-sharing formula based on rewarding productivity rather than stock options for all employees.

Winborn persuaded Chandler to come aboard in an arrangement with Texas A&M for getting his Ph.D. Fate intervened after a couple of months on the job, when Chandler had a malignant melanoma removed with a two-year waiting period for success. By that time the agronomic sales program was so successful that there were nine graduate agronomist, with five Ph.D.s. There was no time for further education. The fertilizer company was so successful that it commanded the attention of Armand Hammer, who was just expanding Occidental Petroleum Company, with Best Fertilizers as his second merger target. The usual merger followed, and all of the merged employees were assured of profit-sharing and uninterrupted employment. The romance lasted a few months, at which point the wily Hammer found a way to abrogate the understanding behind the merger. He trumped the general profit-sharing program that rewards all producers with the then emerging stock option program that rewards only a few at the top. Berry, Winborn and Chandler were fired, thus another lesson in economics. Winborn and Chandler, with Berry's unofficial backing, formed the American Plant Food (APF) Corporation on Berry's principle of agronomic manufacturing and marketing with profit sharing to all. After more than 40 years, APF has turned out to be the most profitable company in the history of the industry.

Pennzoil

"I thought corporate agriculture would last from three to five years. Remember in the late 1960s when corporate agriculture was considered the model for the future? All the gurus on Wall Street, led by Senator George McGovern, were telling big corporations to purchase land. The word was out: We have to feed the world! Pennzoil owned land over much of the United States, actually close to 300,000 acres. They had a 9,000-acre lignite reserve in central Texas for a farming operation.

I was able to get the job as manager. Everyone said that I was over-qualified," confided Chandler.

Scott McWilliams, their University of Purdue, Forester and Operations Land Manager, checked out Chandler and insisted that they hire him for the central Texas operation. The company chose a Washington lobbyist/lawyer to be vice-president of agriculture. This allowed him to pump Chandler's resumé. The company also created a two-headed management group with Foy Phillips, a chemical engineer, to head up Planning and Marketing. Fortunately, he relied on McWilliams and Chandler to operate the Central Texas farm. For two years, Texas came in ahead of income projections and under budget. The reverse was true for their other consortium farms.

"It didn't take three to five years. Pennzoil got into trouble financially with their offshore drilling, and had to have cash. That liquidity crisis caused the firm to sell the farm and lignite reserves after two years. I had the choice of being moved back to Houston or to the West Coast, or accepting several other options. My wife resolved the issue. She told me that I could take any job I wanted, but to make sure I would be happy, as she and the kids were going to stay on the farm. In 1972 I started farming on a few acres, supplementing my income with agricultural consulting. That is when I really got active in productive agriculture with this mid-life career change."

The above is a typical opener for Chandler, the Presenter. He does not apologize for his years with an industry he now calls into question, but he never trembles when he tells audiences that sustainable agriculture is the direction that agriculture has to go.

On to Farming & Consulting

The world has turned over several times since the events described above. In 1970 Chandler left the fertilizer industry for farming and consulting. In 1981, he became the proprietor of the Texas Plant & Soil Lab in Edinburgh, Texas. The Rio Grande Valley and much of rural America has never been the same since then. During the ensuing years, many voices sounded a challenge to high-analysis fertilizers and those opposed to legislate biology. Perhaps most unique of all was the voice of Esper K. Chandler.

Herein are Chandler's words exactly as spoken later to an eco-farming conference. One can detect the easy cadence of his presentation as he tears down conventional ignorance and constructs his logic.

"Being an early student of *Acres U.S.A.* and of the National Organization for Raw Materials since the early 1970s when I left the fertilizer business and moved to the farm, I took a job managing a large Pennzoil United corporate farm, and I was able to pull my kids out of Houston. It required a great sacrifice of income to save our kids from a junior high school with drugs, violence, and lack of control. We moved them to the country, in a superior Rosebud-Lott school system, a heart-of-Texas, small, rural farming community, and I have never regretted it. There were still Christian family values holding sway with a congenial meld of several diverse ethnic groups. This allowed our family to participate in community affairs with leadership roles in religion, school board and functions, and civic service such as Farm Bureau, etc."

The National Organic Standards Board

"You (those that practice sustainable agriculture) are referred to as a subculture," Chandler tells those who are trying to resettle the country for a sustainable future. He counts the subculture remark as an honor. "This nation was founded on its soil. And the subsoil is the base of that soil. The U.S. economic future belongs to the agriculture industry. We have to make choices and decisions about the way we're going. I was fortunate — or unfortunate in terms of the time and personal expense required — in getting appointed to the National Organic Standards Board (NOSB). There is a difference of opinion as to where we are going in natural/organic agriculture. The 1990 Farm Bill contained a section on the OFPA (Organic Food Production Act) to allow a National Organic Standards Board to be formed. This Board was to advise the Secretary of Agriculture. Unfortunately, there was still a lot of resistance to natural/organic agriculture. There was no funding until the spring of 1992. Finally, the powers that be got around to appointing a Board. Several people, including Malcolm Beck of San Antonio, rejected a role for themselves, but insisted that I take a slot. The Board was composed of well-meaning amateurs and dilettantes, as well as qualified, experienced organic producers, processors, and marketers. As a scientist, I gathered a lot of support from the natural/organic community and also from the farm chemical industry. I was selected as the only scientist on the Board."

The National Organic Standards Board worked diligently under USDA's program leader, Harold S. Ricker, and sympathetic staff. They crisscrossed the country holding public hearings and observing local programs. Eugene Kahn, Robert Quinn, Craig Weakley, and other experienced growers, handlers, processors, and marketers sought pragmatic recommendations. Remarkable consensus was achieved among a very diverse membership, but the Board was hamstrung by lack of funds and institutional support. During an election year, Democratic party loyalists blocked George H.W. Bush's substantially increased budget for the board. Only staff manipulations to squeeze funds from other sources allowed the Board to proceed, using conference calls. It was a devastating compromise. No new input arrived now that a new consciousness of Board authority surfaced throughout the country. Previous wide-open procedures allowed almost all germane issues to be thoroughly explored by open inputs until the final votes were cast by the board. Though they originally agreed to operate under *Robert's Rules of Order,* that was soon changed by unanimous approval of Chandler's motion that proceedings be informal with *Robert's Rules* as a guide, rather than a debating society to decide who has the floor, unless contention and debate threatened to demolish the purpose of the rule-making exercise. Although inputs were often hot and heavy, decorum was amazingly respectful. An undertow of suspicions between conservative Republicans and liberal Democrats surfaced to sour the air. There was an undercurrent about Chandler having been named from a Republican state by Secretary of Agriculture Rick Perry, who had defeated populist Jim Hightower with a hefty assist from the Farm Bureau. Chandler rapidly gained acceptance by his insistence on hearing all sides of an issue before votes were cast.

"With my appointment," Chandler reflects, "I rolled back the clock to the days of Louis Bromfield and meetings of Friends of the Land at Malabar Farm. William A. Albrecht, Firman Bear, Cyrus Hopkins and several of the great professors of the Depression era clearly identified the direction that agriculture was required to take. As a child on a worn-out cotton farm in North Louisiana, I read the *Reader's Digest* series by Bromfield and others. After coming in for long lunch breaks for the kids, I got to reading those articles. I had an uncle, Luther Monzingo, who had a job with Gulf Oil during the Depression, and he subscribed. He believed that the family should have up-to-date reading material, for which reason *Reader's Digest, Collier's,* and *Saturday Evening Post* came to our mailbox. Louis Bromfield hit home. I learned that on a run-down farm, the first thing you needed to do was lime it and apply phosphate and then bring back legumes."

Chandler the Consultant

As a private practitioner, Chandler is a generalist, a fact that he is quick to point out to all who will listen. He is less a Will Rogers in humor than in nostalgia. In passages of this book, we hear of his childhood during the Depression and the drift into oil, a euphemism for satellites of oil, namely salt fertilizers. Chandler often tells his audiences about Louis Bromfield and Ruben Douglas of "All's Well Farm," a Gilliam neighbor from whom he had several years of tutelage. Chandler picked up some of what his contemporaries called heresy by reading his uncle's subscriptions. As he speaks, this once redheaded, now graying, gentleman paints a picture worthy of Norman Rockwell's *Saturday Evening Post* covers.

"Once sandy soil, now red clay," he says, "was worn out. So it struck a chord with me as a young person, and as I went through life it stayed with me as I attended Louisiana State University as an agricultural student." When he returned to do graduate work, a professor required him to research J.I. Rodale, a small electronics manufacturer who had picked up on natural systems from Sir Albert Howard, a farm practitioner in India. Was this fact, fancy, or fiction?

The conclusion was inevitable. Organiculture was based on scientific evidence, but it was running against the tide. It was the dawn of the chemical age at the time. The mantra was "Better living through chemistry." Chandler then reminds audiences that his professor required him to spend a great deal of time studying William A. Albrecht, the University of Missouri soil microbiologist who placed in escrow much of the knowledge being summoned today by aficionados of eco-agriculture. Book learning gave way to direct work with the great professor, this near Katy, Texas with a private consultant named Zamzow in tow. Espousing the benefits of lime phosphate rock, Chandler allows that the philosophy of a healthy soil was encoded in him by Albrecht and Rodale. There also were professors who expected to allow the truth to flow from the facts. "The best weren't always the most popular," he adds.

It is to his exposure to the leaders of the grand movement called eco-agriculture that Chandler credits his present insight and outlook. He usually cites contemporaries who have become real innovators: Arden Andersen, Gary Zimmer, Neal Kinsey, Dave Larson, etc. "Are we at the dawn of a new age," asks Chandler, "or are we at the end of an age of productive soils in this nation?" Then he answers his own question. He does not see ruin as the end product of errant farming simply because innovators are exhibiting such resourcefulness and unillusioned self-sufficiency.

Chandler is never slow to credit *Acres U.S.A.* for its leadership, or NORM (National Organization for Raw Materials) for its pioneer work on parity, the exchange equation, and the structural balance that ought to be the legacy of the scholarship in charge.

"The soil is any nation's greatest natural resource, and it is renewable. It creates new wealth. That's the reason this country itself is as productive in agriculture as it has been, but we've neglected the basics. Who, what, when, where, why?" Chandler says, often startling his audiences. In answering these five questions, Chandler points to the soil test. "What does it do?" Strange as it may seem, only a few farmers test their soil.

"The soil test is a chemical way to try to predict what will be available to the plant. But it is only the starting point, only a guide." As he warms to his subject, the presenter inserts his keen observations. "Profitability is what keeps the world going around. We want to get the yield up so we can increase profits. How do we get yield up? By using more inputs. That's the conventional wisdom of chemical agriculture. But folks, we have broken the mold. We can dump more chemicals on the soil, but we don't increase profitability. We don't increase yields or quality. In parts of the Rio Grande Valley in Texas we have burned our soils out." These are harsh words indeed when spoken to professors who defend their bibliographies and to farmers who know no different.

The presenter tells his story the way he feels it. "We switched to drip irrigation. That helped some, but water is only the basis of a complete program. But we have to go back to soil basics so we can move forward." Chandler stresses that he is never after the highest yield, but the yield that lowers cost per unit to the lowest possible point so that general profits are greater. It is up to the grower to determine whether chemical inputs or natural inputs or recycled inputs best serve the objective. "Folklore aside," says Chandler, "we have to replace what we haul off to market. We create that new wealth with higher yield, but we take it off the farm." The answer to the dilemma is the in-depth soil test.

The presenter punctuates his discourse with memorable scenes from a peek into Russia. The terrain looks like the U.S. during the Depression, Chandler recalls. Without many inputs, farmers are reduced to recycling used animal feed, if there is any. The nitrogen cycle can't work, nor can the carbon cycle in such a primitive agriculture, even though red clover was often a primary source of nitrogen in the area visited.

The U.S. is quite different. There was a time when sodium nitrate was just about the only available input. It was spoon-fed. All that changed after World War II. For this reason, and for many collateral reasons, the lessons of sustainable agriculture almost passed from the scene. But they are coming back, Chandler promises.

The presenter does not cover the water input other than to point out that efficient use goes up with balanced fertility. He can't. The range is too great. But he always has the time to tell his listeners that farming will always be on the bottom until farmers lose their fierce independence and work together for a common goal. He tells farmers to get involved on the political side and on the marketing side. Like any good populist, Chandler disdains the commodity trader and the speculators, in short, the crap game that the Board of Exchange and Mercantile Exchanges have become.

Such asides are called for, to be sure, but the discourse swings back to yields and balance as if directed by a magnet. The starting point is the soil test. The seedbed is, after all, a root bed, and the root has to pick up the nutrient and take it into the plant. "Generally, we think of testing the top six inches. But we have to think of taking that mother lode in the top six inches and moving it down into the subsoil," Chandler says. "Think like a root. The root will go down deep if there is anything to go down there for." There is a right way to take a soil sample and a plant sample.

The soil is forever changing. In nature, there are no absolutes. There is only a dynamic system. Always, the presenter has point for point suggestions on sampling, on growing a corn plant, on petiole testing, on cafeteria plant-feeding, and on nurturing a crop the way a mother nurtures a child. Such directions are helpful, but even more helpful is a good history of the soil in question if a consultant is to be of any use.

Standard directions attempt to deal with the topsoil, but Chandler is interested in the second increment: the subsoil. Always sample in-depth by increments. How deep? That depends on the crop. If the crop is forage, the top increment may be somewhat shallow. When regeneration of the soil in the root zone is the objective, the top increment may well be 12 inches.

All of the above is why Chandler wonders aloud whether man's small grasp of natural design has been improved upon! Chandler always reminds his listeners that during the darkest hours of rated fertilizer dominance, the test of the times was hard nitrogen, factory-acidulated fertilizers, and an invisible down-shouting of Nature's way.

Chandler is among the first of the credentialed consultants to enter-
tain Phil Callahan's matchless work, *Paramagnetism: Rediscovering
Nature's Secret Force of Growth*. Foliar fertilization languished after
Sylvan Wittwer's *Non-Root Feeding of Plants* until Chandler and colleagues moved it into the mainstream. T. Senn's work on humates
remained an academic exercise, or a technology used only by a few
producers until

Chandler tied it to drip-irrigation and jump-started not only phosphate uptake, but also the miracle of photosynthesis.

The new natural growth-products that have come on line over the
past 30-40 years, only to be pooh-poohed by land-grant colleges, ignored
legislated formulas and became current coin only because someone
"asked the plant." A century of findings has not been set aside by the
arrival of new knowledge about trace nutrients, even ocean-water nutrients. Rather, these findings have been enhanced. Chandler tells farmers
how major, secondary, minor, and trace nutrients come together.

Cobalt is all but missing from American soils. Yet, nature has decreed
that cobalt governs no less than eight elements in the genetic code:
vanadium, chromium, manganese, iron, nickel, copper, zinc, molybdenum and germanium. The pantheon of minerals covered in *Fertility
from the Ocean Deep* and in *Minerals from the Genetic Code* pries open
a door too long bolted shut by obstinate faculty tied to an almost Stone
Age agronomy, and Chandler never stints in relating the story.

The signal phrase for this discussion of traces is "maximum genetic
potential." When the potential is 20-30 rows of grain, all governed
by the embryo at birth, so to speak, the genetic potential is achieved
by timely availability of nutrients. "All this is possible when we turn
to natural products, hormones, seaweeds, rationed phosphates, and
humic acids," counsels Chandler. "All of these new aids make transfer of
nutrients possible so that nothing can stunt the genetic potential in the
formation stage. What would stunt it? A lack of zinc, a lack of nitrogen,
too much nitrogen, a lack of boron, a lack of calcium . . ."

How to manage the knowledge that Chandler bestows is ever a challenge. "The best way is to pretend that you have a role in Death of a
Salesman. You have to memorize your lines. You have to memorize the
lessons and the basics so that they become second nature." With a measure of humility, Chandler cautions, "Don't ever think you're so right
that you can't be wrong."

"If you are going to grow tree crops — citrus, pecan, peaches, and
so forth — take the second 12 inches and reach on down there to four

feet. Fourth to the fifth foot," he adds. Why? Phosphorus will move in many soils slowly by leaching and root transfer. In some soils it becomes fixed without downward migration. Often nitrogen is based quite deep, a consequence of earlier excessive anhydrous practices. Stored nitrogen can often last for three or four years or more, and calcium can migrate down well out of reach. Other nutrients are often deep-mined, especially potash, calcium and magnesium.

Chandler drops enough hints and inside information to keep a diligent student busy for a month. Take the word "hypoxia" . . . It is an environmental term common in the Mississippi Basin. Waters are supposedly running out of oxygen because of the fertilizer nitrogen from farms in Rainbelt waters. The charge is being made that nitrogen used in the Midwest migrates to Louisiana, Texas, Mississippi, even to Arkansas and on into the Gulf, but little is said about urban lawns and sewage from cities dumped into the rivers of the basin.

Soil samples should take into account spot problems, which should not be deleted. Grid sampling seems to suggest itself, but Chandler opts in favor of representative precise sampling. In the final analysis, the level of management dictates the choice of methods. The ramifications of the sampling art nowadays include global positioning of fields, special isolation of saline-seep areas, and superb management . . . not all of them within the purview of the average well-managed farm.

Using slides, overheads, PowerPoint, and handouts, Chandler often tells his story. "You have green machines with global positioning systems so that when they take that sample in less than a three-foot radius, they install those data into the computer. The jury is still out as to whether massive data better serves the grower than sampling subservient to the natural/organic art."

USDA is now using an aerial viewfinder with remote sensing of the field to determine where the sample spots are going to be. This is the infrared system that can pick out problem areas. Parenthetically, it might be noted that global positioning is now being used to trap insurance farmers who sometimes create crop failure by not planting in the first place. The practice is settling down as the word goes forth explaining how insurance-claim cheaters are being caught. As with any human endeavor, there is always a margin for error. The beautiful profile of the top six inches can be canceled out by a subsoil pretending to be a topsoil, all without the farmer or consultant knowing it.

Chandler closes on a note that all farmers understand. "First you make a crop." It's the lingo of the Cajun cook making gumbo, "First you make the rue (gravy)." It takes organic matter. This consideration transfers itself into the lab, as explained earlier. It's the concept of heating and harsh grinding of soil samples that bears reiteration. The pH is raised two and three points by heating the sample, and the process changes the soil's magnetism and cation exchange capacity. When Chandler finishes his presentation, he is seldom allowed to escape. Trapped by the new converts, he is required to stay on for answers.

Chandler's Family & Associates

Chandler knows that his evolution from a rural environment was nurtured by many accomplished leaders whom he met in education, business, public service, religion, the military, industry, research, and agriculture. They took the time to nurture his desire to be the best that he could be. His progress through life was exceptional because of the many opportunities he had to explore the world. His greatest blessing turned out to be his wife, Bettie Brownlee, a soul-mate, a mother, business manager, and public service leader in her own right. Two Aggie sons, Clyde and Mike, have followed in his footsteps with agronomy degrees to help their father to establish his consulting lab and to operate the family farm while pursuing their own careers. Success has been achieved by both. Mike is a turf and golf course superintendent. Associates claim that he can grow grass anywhere. Clyde is a farmer-turned-lawyer who says that he can now afford to cattle farm the way he wants to. Daughter Jeni has an exceptional personality and many talents, too.

We cannot distil the sum of Chandler's wisdom in a single chapter or even in a single book. But we can mine his thoughts for the nuggets of truth that they contain and for the outline that agriculture has to follow if it is to rescue itself from cumulative blunders. Chandler sometimes jokingly describes himself as a part of the mañana generation, a reverse metaphor if ever there was one. He usually opens the laboratory at 6:30 a.m., or at 7 a.m. if he sleeps late. He closes it after everyone has gone home. Levity aside, there is no waiting until tomorrow for this tireless scientist and farm operator.

(Publisher's note: Esper K. Chandler died after approving the final draft of this book, but before its publication.)

Appendix B

Texas Plant & Soil Lab

The first soil-testing laboratory in the State of Texas has now celebrated 70 years of service to growers. Now entering their third generation of ownership, with the passing of Esper K. Chandler, The Texas Plant & Soil Lab, Inc. of Edinburg, Texas continues to be internationally recognized as a leader in practical agricultural consulting. The Lab furnishes soil, water, and plant analyses that are a uniquely more accurate method of soil testing with easy-to-understand interpretations and recommendations. Also noted as leaders in soil fertility and plant nutrition for most crops as well as grasses used as turf on athletic fields, golf courses and especially forage crops.

Contact Information

Texas Plant & Soil Lab, Inc.

5115 West Monte Cristo Road

Edinburg, Texas 78541

Phone: (956) 383-0739

Fax: (956) 383-0730

Website: www.tpsl.biz

Email: info@tpsl.biz

Physical Address: FM 1925, 3.5 miles west of US 281

Index

2,4-D, 238-240

Acres U.S.A., 121, 224, 249, 252
Albrecht Papers, The, 238
Albrecht, William A., 2, 41, 55, 84, 88, 102, 128, 139, 205, 207, 234-238, 250-251
Alzheimer's disease, 224, 226
American Journal of Industrial Medicine, 226
American Plant Food Corporation (APF), 247
ammonia, anhydrous, 99, 240, 244
ammonia, 85, 86
ammonium nitrate, 85
Anderson, F.I., 102
Andrews, W.B., 99, 244
application rate, 111-113
"Ask the Plant", 17, 61, 184
Association of American Plant Food Control Officials (AAPFCO), 149
Atomic Energy Commission, 29

Backster, Clive, xiii, xiv,
bananas, 198
bat guano, 85-86
beans, 67
Beck, Malcolm, vii, xv, xvii, 125, 226, 228, 229, 241
beef, grass-fed, 219
Beijerinck, Martinus, 99
Bermudagrass, 215-216, 244-245

Bermudagrass, coastal, 118
Bermudagrass, nitrogen requirements for, 244-245
Berry, Lowell, 247
blueberries, 68
Bordeaux mixture, 142
Borders, Nowell, 190
boron, 5
boron, deficiency of, 12
Bradford, Joe M., xiv, 40134
brassicas, 67
Brix reading, 20
Bromfield, Louis, 234, 235-236, 250
BST, 115
Burton, Glenn, 118
Byron, Lord, 27

calcium, 5-6, 8, 16, 45, 72, 79, 81-83, 158-159, 161, 163-165, 217
calcium, deficiency of, 7, 13
calcium phosphate, 24
Callahan, Phil, 125, 254
Calvin, Melvin, 33-36
carbohydrate functions, 136-138
carbon, 33-34, 203
carbon dioxide, 44-45
carbonates, 72
carbon-water cycle, 39
Carrots Love Tomatoes, 197
carrots, 11
Carson, Rachel, 115, 246

cation exchange capacity (CEC), 51, 56, 60, 70, 139

celery, 67

cereal grains, 69

Chandler, Bettie, 246

Chandler, Esper K., vii, xiv, 2, 99, 119, 140, 205, 223, 233-256

Chandler-Lengyel system, 55

chlorophyll, 28

Christy, David, 204

citrus, 48, 68, 209-214

citrus, leaf analysis, 211

citrus, nutrient requirements of, 210-213

citrus, stages of growth, 210

Clement, H.F., 139, 201

CO₂ method, 55-57, 70, 122, 141

coastal Bermudagrass, 118

coastal hay, soil analysis example, 222

cobalt, 254

coffee, 199

coffee, production of, 14

Coleman, Russell, 245

compost, 167

compost tea, 128, 220

conservation tillage, 41

copper, 5

copper, deficiency of, 9

corn, 69, 169-174

corn, increasing yields of, 180

corn, magnesium deficiency in, 172

corn, maximum economic yields, 171

corn, nitrogen deficiency in, 172

corn, nutrient requirements of, 173

corn, phosphate deficiency in, 172

corn, plant stages, 170

corn, plant testing, 170

corn, potash deficiency in, 172

corn, roots, 53

cotton, 15, 31, 67, 204, 206

Cotton is King, 204

crape myrtle, 124

Crop Nutrient Needs, 59

crop-logging, 60, 139

Cumming, Jim, 229

Davis, Archie, 15

Davis, Walt, 215

Davy, Humphry, 96

DNA, 28, 184-186

Do Pesticides Cause Lymphoma?, 226

dolomite lime, 82

Draper, Kathleen, 135

drip-irrigation, 16, 25, 80, 93, 101, 184, 190-191, 196, 223, 254

dung beetles, 215

E.C., 160, 162

Earp-Thomas, George H., 135

Economic Report of the Producers, 205

Einstein, Albert, 19

electrical conductivity (EC), 71-72

Elements of Agricultural Chemistry, 96

Farmer of Tomorrow, The, 102

Farmers of Forty Centuries, 103

Faulkner, Ed, 35

Fertility From the Ocean Deep, 20, 228, 254

Fertilizer Institute, 104

fertilizer, 92, 98, 105

fertilizers and water use, 37

Fisher, Flake, 176

forages, 68

Frankenstein, 27

From My Experience, 236

fruit trees, 68

Garcia, Noel, 191
Garden-Ville, xiv
Garrett, Howard, 123, 125
Genesis, 19
glyphosate, 228-230
GMO, 115, 184, 186-187
golf courses, 226, 227
Graff, Ken, 221
grains, 182
grains, cereal, 69
grapes, 68
Grew, Nehemiah, 95-96
guano, bat, 85-86
gypsum, 16, 81, 98

Hales, Stephen, 96
Hammer, Armand, 85, 92
hardpan, 160
Harmel, Daren, 114
Harr, Thomas, 44
hay, 68
Hippocrates, 95
Holistic Resources Management, 220
Howard, Albert, 95, 140, 143, 146, 251
humic acid, 93, 157
humus, 166, 222
hypoxia, 255

Ingalls, John J., 221, 224
Ingham, Elaine, 128, 135
International Journal of Parapsychology, xiv
iron, 5
iron, deficiency of, 9-10

Jansen, Don, 20

Johns, Dawson, 117
Journal of Occupational and Environmental Medicine, 227

Kika de la Garza USDA Research Station, xiv, 40, 101
King, F. H., 103
King, James "Bubba", 2, 190
K-Mag, 84
Kneese, Melinda, 135
Korean War, 241
Krebs cycle, 27, 34, 36, 176
Krebs, Hans Adolf, 34-35
Kyle, Thomas, 190

Larson, Dave, 100
Lawe, Robert, 3
LD$_{50}$, 228
leaf monitoring, 62
leaf/petiole analysis, 23
leaf/petiole sampling, 21-22
legumes, 68
Lengyel, Albin D., 3, 14-15, 47, 139-140.
Lesikar, Bruce, 114
lime, 91
lime, dolomite, 82
Logan, Dan, 16, 42, 238-240, 242
Lou Gehrig disease (ALS), 224, 226
Louisiana State University Experimental Station, 216, 221, 234-235, 242, 244

magnesium, 5, 45, 72, 83, 158-159
magnesium, deficiency of, 13
Malabar Farm, 234-236, 250
manganese, 5
manganese, deficiency of, 10-11
Manual on Phosphates in Agriculture, 97

manure, 44

maximum economic yield (MEY), 180

McFarland, Mark, 59

Mechell, Justin, 114

Medina, 45, 100, 147, 149

melons, 67

Mendeleev, xiv

Mendeleev table, 28

Meyerhof, Otto, 34

microbes, 92, 146, 217, 225

micronutrients, 30, 44, 107, 128, 135, 153, 192, 225

Mineralization, Will it Reach Us In Time, 234

Minerals from the Genetic Code, 254

MK Labs, 221

molasses, 135-136

molybdenum, 91

Monsanto, 121, 185

Morgan, Nevin D., 145

Murray, Maynard, 20

National Organic Standards Board (NOSB), 249-250

National Organization for Raw Materials (NORM), 252

National Plant Food Institute, 104, 144, 245

nitrate, 5, 24, 45, 70-71, 154-156, 166

nitrogen, 6, 85, 108, 152, 156, 177, 179, 97-98, 107, 127,

nitrogen, deficiency of, 7-8

nitrogen, fixation of, 99

Non-Root Feeding of Plants, The, 147, 254

NPK, 90-91, 94, 98, 104-105, 110, 143, 195

Nutrient Monitoring Program, 61, 63-67

ocean minerals, 93

oleander, 214

onions, 182

onion, nutrient requirements of, 183

Oppenheimer, Carl, 149

Organic Food Protection Act, 41

organic matter (OM), 39-40, 70, 110

Ozolins, Rudolf, 106

paramagnetism, 125, 202

Paramagnetism: Rediscovering Nature's Secret Force of Growth, 254

pastures, 215-224

pastures and cattle, 220

peanuts, 206-208

peanuts, calcium and, 206, 208

peanuts, magnesium and, 207

peanuts, phosphate and, 207

peas, 67

pecans, 68, 129-134

Pennington, Dale, 216

Pennzoil, 247, 248

peppers, 67

petiole monitoring, 62

petiole nutrient analysis, 192

petiole testing, 60, 66, 194

Petit, Bob, 122, 145, 176, 246

Pfeiffer, Ehrenfried E., 105, 126

pH, 70, 142, 199

Phillips 66 Petroleum Company, 105, 243

Philodendron, xiv

phosphate, 5-6, 24, 45, 70-71, 86-90, 97, 101, 105, 127, 152, 154-157

phosphorus, 45, 58-59, 85, 87, 127, 176, 179, 199

phosphorus, deficiency of, 7-9, 11

photosynthesis, 28, 35, 39

Pickard, Barbara, xiii

plant analysis guide sheet, 5
Plant Analysis Handbook II, 32
plant analysis testing, 61
plant nutrient process, 176
plant nutrition monitoring, 61
Plowman's Folly, 35
potash, 85, 99, 104, 106, 108
potassium, 5, 12, 70, 72, 83, 85, 158-159, 164-166
potassium, deficiency of, 11
potatoes, 12, 67, 196
Pratt, Neal J., 216
Priestly, Joseph, 96

raspberries, 68
Raw Material Economics, 205
refractometer, 19-20, 23
Rio Grande Valley, xiv, xvii, 82, 148
Riotte, Louise, 197
Rodale, J.I., 140, 146, 234, 251,
roots, corn, 53
Roses Love Garlic, 197
Ross, Betsy, 220
Roundup, 229
Rowett Research Institute, 185
Russell, Darrell, 117
Russell, Walter, 28

salt, 44, 73-75, 78, 160, 162, 166-167
salt cations, 71
Savory, Allan, 220
scandium, 28
Schultz, George, 47-48, 56
selenium, 92
Senn, T., 78, 122, 145
Shelley, Mary, 27
Silent Spring, 115, 246
sodium, 5, 45, 73, 79, 163-165, 167
Soil Bank, 118

soil samples, 49
soil testing, 49-52, 70, 73-74, 76-77
soil tests, types of methods, 78
soil, texture of, 73
soil tilth, 93, 160
sorghum, 169
soybeans, 12, 67
sports fields, assessing nutrients in, 225
Stall, Arthur, 34
Steiner, Rudolf, 44
Stichler, Charles, 59
strawberries, 68
Sturgis, M.B., 234
Stutte, Charles A., 15
subsoil, 76
sugarcane, 69, 200-204
Sugarcane Crop Logging and Crop Control: Principles and Practices, 139, 201
sugars, 136-138, 203
sulfur, 16
Sul-Po-Mag, 84
sunflowers, 68
superphosphate, 16, 97, 106

Tambora volcano, 27
Tennessee Valley Authority (TVA), 39-40, 99, 147
TerraClean, 189
Texas A&M University, 57, 147, 216, 246-247
Texas Agricultural Extension Service, 59
Texas Plant & Soil Lab, 47, 70, 121, 141, 184, 194, 206, 217, 248, 257
texture, soil, 73
Thaer, Albrecht Daniel, 3
tilth, soil, 160
Tisdale, Samuel, 245

tomatoes, 67
tree diagnosis, 123
tree root flares, 124
Tuning into Nature, 125
turf, 68
turfgrass, 225-227

Unforgiven, 237

urea, 85-86
van Helmont, Jan Baptista, 95-96
vitamin D, 28
von Baeyer, Adolf, 34
von Liebig, Justus, 26, 34, 41, 96, 98, 103, 216, 231

Walters, Charles, xvii, 205
water, 25, 36
watermelon, 189-196
watermelon, drip-irrigation, 195
Way, Thomas, 103
Weeds: Control Without Poisons, 4, 199
wheat, 176-179

wheat, nitrogen requirements of, 178
wheat, stages, 175
wheat, increasing yields of, 180
wheat, nutrient requirements of, 175, 178
wheat, phosphorus requirements of, 178
Wills, Harold, 29
Willstatter, Richard, 34
Wilson, Edward O., 177
Winogradsky, Serge N., 99
Wittwer, Sylvan, 147, 254
World War I, 103
World War II, 40, 85, 96-97, 143, 234, 241, 253
Worley, B., 109
Wynd, F. Lyle, 107

Young, Ray, 42

Zibilske, Larry M., 134
zinc, 5
zinc, deficiency of, 10

Also from Acres U.S.A.

Eco-Farm: An Acres U.S.A. Primer
BY CHARLES WALTERS

In this book, eco-agriculture is explained — from the tiniest molecular building blocks to managing the soil — in terminology that not only makes the subject easy to learn, but vibrantly alive. Sections on NP&K, cation exchange capacity, composting, Brix, soil life, and more! *Eco-Farm* truly delivers a complete education in soils, crops, and weed and insect control. This should be the first book read by everyone beginning in eco-agriculture . . . and the most shop-worn book on the shelf of the most experienced. *Softcover, 476 pages. ISBN 978-0-911311-74-7*

Weeds: Control Without Poisons
BY CHARLES WALTERS

For a thorough understanding of the conditions that produce certain weeds, you simply can't find a better source than this one — certainly not one as entertaining, as full of anecdotes and homespun common sense. It contains a lifetime of collected wisdom that teaches us how to understand and thereby control the growth of countless weed species, as well as why there is an absolute necessity for a more holistic, eco-centered perspective in agriculture today. Contains specifics on a hundred weeds, why they grow, what soil conditions spur them on or stop them, what they say about your soil, and how to control them without the obscene presence of poisons, all cross-referenced by scientific and various common names, and a new pictorial glossary. *Softcover, 352 pages. ISBN 978-0-911311-58-7*

Science in Agriculture
BY ARDEN B. ANDERSEN, PH.D., D.O.

By ignoring the truth, ag-chemical enthusiasts are able to claim that pesticides and herbicides are necessary to feed the world. But science points out that low-to-mediocre crop production, weed, disease, and insect pressures are all symptoms of nutritional imbalances and inadequacies in the soil. The progressive farmer who knows this can grow bountiful, disease- and pest-free commodities without the use of toxic chemicals. A concise recap of the main schools of thought that make up eco-agriculture — all clearly explained. Both farmer and professional consultant will benefit from this important work. *Softcover, 376 pages. ISBN 978-0-911311-35-8*

To order call 1-800-355-5313
or order online at www.acresusa.com

Hands-On Agronomy

BY NEAL KINSEY & CHARLES WALTERS

 The soil is more than just a substrate that anchors crops in place. An ecologically balanced soil system is essential for maintaining healthy crops. This is a comprehensive manual on soil management. The "whats and whys" of micronutrients, earthworms, soil drainage, tilth, soil structure and organic matter are explained in detail. Kinsey shows us how working with the soil produces healthier crops with a higher yield. True hands-on advice that consultants charge thousands for every day. Revised, third edition. *Softcover, 352 pages. ISBN 978-0-911311-59-4*

Hands-On Agronomy Video Workshop
Video Workshop

BY NEAL KINSEY

 Neal Kinsey teaches a sophisticated, easy-to-live-with system of fertility management that focuses on balance, not merely quantity of fertility elements. It works in a variety of soils and crops, both conventional and organic. In sharp contrast to the current methods only using N-P-K and pH and viewing soil only as a physical support media for plants, the basis of all his teachings are to feed the soil, and let the soil feed the plant. The Albrecht system of soils is covered, along with how to properly test your soil and interpret the results. *80 minutes.*

The Biological Farmer
A Complete Guide to the Sustainable & Profitable Biological System of Farming

BY GARY F. ZIMMER

 Biological farmers work with nature, feeding soil life, balancing soil minerals, and tilling soils with a purpose. The methods they apply involve a unique system of beliefs, observations and guidelines that result in increased production and profit. This practical how-to guide elucidates their methods and will help you make farming fun and profitable. *The Biological Farmer* is the farming consultant's bible. It schools the interested grower in methods of maintaining a balanced, healthy soil that promises greater productivity at lower costs, and it covers some of the pitfalls of conventional farming practices. Zimmer knows how to make responsible farming work. His extensive knowledge of biological farming and consulting experience come through in this complete, practical guide to making farming fun and profitable. *Softcover, 352 pages. ISBN 978-0-911311-62-4*

To order call 1-800-355-5313 or order online at www.acresusa.com

Agriculture in Transition

BY DONALD L. SCHRIEFER

Now you can tap the source of many of agriculture's most popular progressive farming tools. Ideas now commonplace in the industry, such as "crop and soil weatherproofing," the "row support system," and the "tillage commandments," exemplify the practicality of the soil/root maintenance program that serves as the foundation for Schriefer's highly-successful "systems approach" farming. A veteran teacher, lecturer and writer, Schriefer's ideas are clear, straightforward, and practical. *Softcover, 238 pages. ISBN 978-0-911311-61-7*

From the Soil Up

BY DONALD L. SCHRIEFER

The farmer's role is to conduct the symphony of plants and soil. In this book, learn how to coax the most out of your plants by providing the best soil and removing all yield-limiting factors. Schriefer is best known for his "systems" approach to tillage and soil fertility, which is detailed here. Managing soil aeration, water, and residue decay are covered, as well as ridge planting systems, guidelines for cultivating row crops, and managing soil fertility. Develop your own soil fertility system for long-term productivity. *Softcover, 274 pages. ISBN 978-0-911311-63-1*

The Non-Toxic Farming Handbook

BY PHILIP A. WHEELER, PH.D. & RONALD B. WARD

In this readable, easy-to-understand handbook the authors successfully integrate the diverse techniques and technologies of classical organic farming, Albrecht-style soil fertility balancing, Reams-method soil and plant testing and analysis, and other alternative technologies applicable to commercial-scale agriculture. By understanding all of the available non-toxic tools and when they are effective, you will be able to react to your specific situation and growing conditions. Covers fertility inputs, in-the-field testing, foliar feeding, and more. The result of a lifetime of eco-consulting. *Softcover, 236 pages. ISBN 978-0-911311-56-3*

Bread from Stones

BY JULIUS HENSEL

This book was the first work to attack Von Liebig's salt fertilizer thesis, and it stands as valid today as when first written over 100 years ago. Conventional agriculture is still operating under misconceptions disproved so eloquently by Hensel so long ago. In addition to the classic text, comments by John Hamaker and Phil Callahan add meaning to the body of the book. Many who stand on the shoulders of this giant have yet to acknowledge Hensel. A true classic of agriculture. *Softcover, 102 pages. ISBN 978-0-911311-30-3*

Alternative Treatments for Ruminant Animals

BY PAUL DETTLOFF, D.V.M.

Drawing on 36 years of veterinary practice, Dr. Paul Dettloff presents an natural, sustainable approach to ruminant health. Copiously illustrated chapters "break down" the animal into its interrelated biological systems: digestive, reproductive, respiratory, circulatory, musculoskeletal and more. Also includes a chapter on nosodes, with vaccination programs for dairy cattle, sheep and goats. An information-packed manual from a renowned vet and educator. *Softcover, 260 pages. ISBN 0-911311-77-7*

Grass, the Forgiveness of Nature

Exploring the miracle of grass, pastures & grassland farming

BY CHARLES WALTERS

In this wide-ranging survey of grass forages and pastureland, Charles Walters makes the case that grass is not just for cows and horses — that in fact it is the most nutritious food produced by nature, as well as the ultimate soil conditioner. You will learn from traditional graziers who draw on centuries of wisdom to create beautiful, lush, sustainable pastures, as well as cutting-edge innovators who are using such methods as biodynamics and sea-solids fertilization to create some of the healthiest grasslands in the world. Leading agronomists not only explain the importance of grasses in our environment, they also share practical knowledge such as when to look for peak levels of nutrition within the growing cycle and how to use grass to restore soil to optimum health. A must-read for anyone interested in sustainable, bio-correct agriculture, this information-packed volume is a comprehensive look at an essential family of plants. *Softcover, 320 pages. ISBN 0-911311-89-0*

Soil, Grass & Cancer

BY ANDRÉ VOISIN

Almost a half-century ago, André Voisin had already grasped the importance of the subterranean world. He mapped the elements of the soil and their effects on plants, and ultimately, animal and human life as well. He saw the hidden danger in the gross oversimplification of fertilization practices that use harsh chemicals and ignore the delicate balance of trace minerals and nutrients in the soil. With a volume of meticulously researched information, Voisin issues a call to agricultural scientists, veterinarians, dietitians and intelligent farmers to stand up and acknowledge the responsibilities they bear in the matter of public health. He writes as well to the alarmed consumer of agricultural products, hoping to spread the knowledge of the possibilities of protective medicine — part of a concerted attempt to remove the causes of ill health, disease and, in particular, cancer. *Softcover, 368 pages. ISBN 0-911311-64-5*

Fertility Farming

BY NEWMAN TURNER

Fertility Farming explores an approach to farming that makes minimal use of plowing, eschews chemical fertilizers and pesticides, and emphasizes soil fertility via crop rotation, composting, cover cropping and manure application.

Turner holds that the foundation of the effectiveness of nature's husbandry is a fertile soil — and the measure of a fertile soil is its content of organic matter, ultimately, its *humus*. Upon a basis of humus, nature builds a complete structure of healthy life — without need for disease control of any kind. In fact, disease treatment is unnecessary in nature, as disease is the outcome of the unbalancing or perversion of the natural order — and serves as a warning that something is wrong. The avoidance of disease is therefore the simple practice of natural law. Much more than theory, this book was written to serve as a practical guide for farmers. Turner's advice for building a productive, profitable organic farming system rings as true today as it did sixty years ago when it was written. *Softcover, 272 pages. ISBN 978-1-601730-09-1*

Herdsmanship

BY NEWMAN TURNER

In this book, Turner explains that livestock illness is a result of bad farming practices and that real livestock health begins with true natural farming disciplines such as composting, biodiverse pastures with deep-rooted forages and herbs, and sub-soiling, as well as the avoidance of supposed panaceas that ignore or marginalize these fundamentals such as vaccines, pesticides, antibiotics and artificial fertilizers. He teaches that the cornerstones of profitability are rooted in herd health, which in turn is rooted in: soil fertility and animal nutrition, cattle breeding for better feed efficiency, and cattle breeding for longevity. Longevity, he holds, is the most critical factor for success in livestock breeding and production. *Softcover, 272 pages. ISBN 978-1-601730-10-7*

Fertility Pastures

BY NEWMAN TURNER

In *Fertility Pastures*, Turner details his methods of intensive pasture-based production of beef and dairy cows in a practical guide to profitable, labor-saving livestock production. He developed a system of complex "herbal ley mixtures," or blends of pasture grasses and herbs, with each ingredient chosen to perform an essential function in providing a specific nutrient to the animal or enhancing the fertility of the soil. He explains his methods of cultivation, seeding and management. There are also chapters on year-round grazing, making silage for self-feeding, protein from forage crops, and pastures for pigs and poultry. He also details the roles individual herbs play in the prevention and treatment of disease. *Softcover, 224 pages. ISBN 978-1-601730-11-4*

To order call 1-800-355-5313 or order online at www.acresusa.com